"Jim Edwards's captivating book [*To ...*] ... gious thought with intellectual proper... is novel. He combines inventor stories with legal and religious principles that seem well matched, adding momentum to his presentation. He ends with a plea for policymakers to revive patent law in America as the foundation of our prosperity today as at the early decades of our history."
—*Hon. Paul R. Michel, chief judge (ret.), U.S. Court of Appeals for the Federal Circuit*

"Americans love technology and the practical benefits it brings. Jim Edwards's book assesses what underlies new innovations—creative or inventive abilities and the inherent right to own what one's efforts produce. This book demonstrates the necessity that creativity and ownership be linked. Jim, who staffed IP issues when I served on the House Judiciary Committee, draws this essential principle from Judeo-Christian Scripture. He describes its purest application in the American patent system as the Founders intended it. *To Invent Is Divine* points to how we can restore that biblical model and boost human flourishing."
—*Hon. Ed Bryant, former member, U.S. House of Representatives (R-TN), magistrate judge (ret.), U.S. District Court for the Western District of Tennessee, and former U.S. attorney for the Western District of Tennessee*

"Very few understand intellectual property. Fewer still could show America's success as 'one nation under God' being built on the bedrock of patents. I thank Jim for this book. As we now see a nation and an economy in need of rebirth, we should look no further than *To Invent Is Divine* to see what God did through our Founders and what he can do through us to restore it now."
—*Jonathan Rogers, chief operating officer, Centripetal*

"Brilliant James Edwards explores how biblical wisdom is a fount of every blessing. Made in his image, the Creator created us to create. We find joy as we promote the flourishing of people, nations, and all of creation. Capitalism honors and incentivizes helpfulness to our fellow man, and voluntary generosity to the vulnerable. Anything less harms people whom God so loves. This book is a key to the renewal of our culture and civilization. 'We are his workmanship, made . . . for good works . . .'"
—*Kelly Monroe Kullberg, general secretary, American Association of Evangelicals (AAE), author,* **Finding God at Harvard***, and founder, the Veritas Forum*

"Public policy discussions of intellectual property rights typically center on purely material concerns, such as the creation, ownership, or distribution of IP-generated wealth. James Edwards's provocative new book, *To Invent Is Divine*, transcends the typical narrative by highlighting the Judeo-Christian understanding that the 'creative human spark' behind IP comes from God—the Creator of man and the Author of all that is good. Seeing IP from this perspective sheds further light on the worth of IP, not just as the wellspring of economic benefits for humans, but as a manifestation of applied God-given talent that should be encouraged to thrive."
—*Alden F. Abbott, senior research fellow, Mercatus Center, George Mason University, and former general counsel, Federal Trade Commission*

"*To Invent Is Divine* is a tour de force of theology, political theory, intellectual history, and intellectual property law. Its description of the historical development and legal details of the U.S. patent system is especially impressive. Jim Edwards clearly conveys why the Founding Fathers empowered the federal government to secure patents and how early American policymakers, judges, and lawyers implemented their vision as a key element of American exceptionalism. Edwards brilliantly makes the case that the success of America as the shining city on the hill is due in large part to the uniquely American patent system."

—*Adam Mossoff, professor of law, Antonin Scalia Law School, George Mason University*

"In this insightful and thought-provoking work, Edwards weaves together stories of inventors and creators, exploring their endeavors through the lens of biblical creation themes. He connects innovation to the divine spark instilled by our Creator, asserting that ownership of creation is a natural extension of dominion. Linking the U.S. patent system to the philosophical foundations of Judeo-Christian spiritual tradition, Edwards highlights the principles of 'as ye sow, so shall ye reap.' He also addresses contemporary challenges to the U.S. intellectual property system, offering a compelling call to safeguard the innovation ecosystem."

—*Laura A. Peter, former deputy under secretary of Commerce for Intellectual Property and deputy director of the United States Patent and Trademark Office*

"Early in his career, Dr. James Edwards was a legislative aide in my Washington office. He is intelligent, kind, hardworking, and simply one of the finest men I have ever known. Now he has written an important new book, *To Invent Is Divine*. I hope it is read by many thousands. The very important relationship between our God-given abilities and property rights has never been adequately recognized. Most young people have not been taught about how essential private property is to both our freedom and our prosperity. Jim Edwards's new book fills a great hole in this area and could help give this nation a brighter and more prosperous future."

—*Hon. John J. Duncan, Jr., member (ret.), U.S. House of Representatives (R-TN)*

"So many scientists, engineers, and inventors spend countless hours devoting themselves to solving extraordinarily difficult problems for the benefit of people they will never meet. Unfortunately, this story has become overtaken by a false narrative that conflates the giving nature of innovators with greed. *To Invent Is Divine* cuts through the policy propaganda and explains how American patent policy is increasingly at odds with, and even belligerent toward, the very people who spend their lives trying to help society. We all benefit from the Herculean efforts of inventors who make life better, and sometimes even possible. Edwards is precisely accurate when he suggests we should have policies and laws that enable those who work tirelessly to solve the biggest problems of the day to be able to spend as much time as possible sharing their talents through inventing a better world for everyone."

—*Gene Quinn, president & CEO, IPWatchdog, Inc.*

"James Edwards masterfully makes the case that humans were created to be inventive—just like our Maker. When we seek to innovate, we are likely to prosper ourselves and others. More importantly, our spirits flourish. Edwards demonstrates the importance of legal incentives to do one of the very things for which we were created. To invent is truly divine!"
—*Bill Wichterman, board president, Faith and Law, and former special assistant to President George W. Bush*

"*To Invent Is Divine* integrates the Bible, American history, principles foundational to human beings, their application in creative endeavors in pursuit of human flourishing, and more. Jim Edwards packs breadth and depth of benefit to a wide audience—everyone, no matter how knowledgeable about any one facet, will learn from this book."
—*Timothy S. Goeglein, vice president for Government and External Relations, Focus on the Family*

"Jim Edwards isn't an academic theoretician but someone who's been actively involved in protecting and promoting our innovation system, which made the United States of America the most innovative country in history. However, that can quickly erode if we don't understand the foundations of that system. In *To Invent Is Divine*, Jim provides important insights into how and why our system was created, which is well worth considering."
—*Joseph P. Allen, executive director, Bayh-Dole Coalition, former professional staff, Senate Judiciary Committee, to Sen. Birch Bayh (D-IN), and founding director, Office of Technology Commercialization, U.S. Department of Commerce*

"Jim Edwards is a master at telling the amazing story of U.S. patent law, articulating how the incentives in our patent system unleashed the creativity of 'ordinary' Americans. Not only did the right to patent extend to groups who did not have the right to vote, including women, free blacks, and immigrants, it led to unprecedented economic growth and flourishing. Jim's genius is in extending the biblical theology of Common Grace to the gifts of creativity and ownership. At the Cade Museum for Creativity and Invention, we had to invent a word to describe the magic that happens when ordinary people invent extraordinary things. We call it 'inventivity.' It was an absolute pleasure to read this book, where inventivity resonates off of every page."
—*Phoebe Cade Miles, cofounder and chairman of the Board, Cade Museum for Creativity and Invention*

"Jim Edwards addresses an existential issue in this probing examination of human achievement. All that we have and all that we participate in creating—intellectual or technological—any form of human value—derive as a gift from God which we hold 'in trust.' As humans we are finite instruments in God's infinite realm. But in our quest for worldly acclaim, we all too often ignore this fact, and it causes consternation among many. We owe Jim a great debt for revisiting the old question: How do we prioritize our human quests for success and achievement?"
—*Hon. John L. Napier, former member, U.S. House of Representatives (R-SC), and judge (ret.), U.S. Court of Federal Claims*

"Jim Edwards provides a warm, welcoming, and thoughtful guide through Judeo-Christian texts and practices for studying the theory and practice of how legal rules about intellectual property can serve as tools for securing ownership in ways that foster creativity and innovation to the betterment of all society."
—*Hon. F. Scott Kieff, Stevenson Bernard professor, George Washington University Law School, and former commissioner, U.S. International Trade Commission*

"One need not be a person of faith to admire faith in others and how it can inspire them to good works. Jim Edwards has crafted a book derived from his deep faith in which he argues persuasively how faith and philosophy led our nation's Founders to craft a patent system available to anyone inspired to invent, believing that would best help the young country develop and compete with the Old World. He explains how that egalitarian system served our country so well for so long, allowing innovative upstarts to challenge and even supplant industry incumbents, just as America's economy challenged and ultimately surpassed those in Europe. He also details the many ways the system has come under strain, as policymakers and courts have been seduced by the unprecedented concentration of money and power in a handful of companies that have actively worked in complicity with policymakers to undermine the patent system to preserve their incumbency. Jim's prescriptions for getting us back on track make this unique book a must-read for anyone who cares about the future of the U.S. economy."
—*Brian Pomper, executive director, Innovation Alliance*

"Made in the image of the Creator, humankind reflects the Maker's extraordinary gifts of creativity. Few individuals of our time have done more than James R. Edwards Jr. to celebrate the spirit of human ingenuity and champion protections for the use and control of the fruits of imaginative, artistic, and inventive minds. He has thought deeply about the theological, legal, and economic aspects of facilitating and protecting human creativity."
—*Daniel L. Dreisbach, professor of Legal Studies, American University, and author,* **Reading the Bible with the Founding Fathers (Oxford)**

"In *To Invent Is Divine*, Edwards brings together a scriptural account of creativity with a general Lockean account of property rights, which is essentially the position of the American Founders. Lincoln captured this creativity-ownership combination when he said the U.S. patent system 'added the fuel of interest to the fire of genius, in the discovery and production of new and useful things.' *To Invent Is Divine* also illustrates the practical benefits that follow from this model, as well as the ills that come when creativity and ownership are separated or rendered tentative. I recommend this book as a caution and a roadmap."
—*Matthew Spalding, dean, Van Andel Graduate School of Government, Hillsdale College Washington, DC, Campus*

"As an inventor and entrepreneur, I often encounter numerous challenges in bringing new inventions to life and delivering them to the patients who need them. Jim's book has been a source of great inspiration for me, as it connects my faith in God with creativity and the stewardship of intellectual property. *To Invent Is Divine* excellently contextualizes the intellectual property system, particularly patents, through the lens of the Bible, highlighting the government's role in implementing biblical principles for the benefit of humanity. I highly recommend this book to everyone. It serves as a powerful encouragement and a reminder of what we can do to rejuvenate our innovation ecosystem for the betterment of society and the glory of God."

—Eb Bright, president and general counsel, ExploraMed

"In this valuable work, Jim Edwards explains that human ingenuity is a gift of our Creator, an expression of his care for those he has made in his image. Creativity combined with the amazing resources God has entrusted to us has led to advances in all spheres of life, and Edwards joins acute theological understanding with a fascinating review of history and technology to show how these advances have benefited so many. A sterling and important book."

—Robert F. Schwarzwalder Jr., PhD, senior lecturer, Honors College & Dept. of General Education, Regent University

"*To Invent Is Divine* is remarkable for its usefulness and its faithfulness. Jim Edwards brings a special set of skills to this book: a man of deep faith and a sophisticated practitioner of policy and lawmaking, especially regarding patents and property rights. I rely on him for guidance in both areas of his expertise. Read *To Invent Is Divine* and pray about it too!"

—Ed Martin, president, Phyllis Schlafly Eagles and Eagle Forum Education and Legal Defense Fund

"I have known Dr. Jim Edwards for several years and consider him a friend and mentor. He is a leading thought leader in intellectual property in Washington, DC, and a man of impeccable character and faith. Jim's new book, *To Invent Is Divine*, highlights the divine inspiration behind invention and creativity, a truth well-known in the intellectual property community. The ancient Hermetics believed that the pursuit of knowledge was the pursuit of the divine, and Jim's book clearly illustrates this. I highly recommend it to anyone interested in intellectual property and its foundational principles."

—Brad Watts, former Republican chief counsel, Senate Judiciary Committee Subcommittee on Intellectual Property

"We serve a God who is creative to His core. The Bible's opening line practically explodes off the page declaring God's instinct to create, and the pinnacle of His creation is a being capable of reflecting this characteristic—you! In *To Invent Is Divine*, James Edwards beautifully captures this truth. You will walk away inspired to live from a place of deep creativity. Why? Because it brings glory to the Creator who endowed you with the capacity to invent, build, and create!"

—Thann Bennett, author, The Equipped *newsletter,* My Fame, His Fame, *and* In Search of the King, *and president of A Fearless Life*

"Edwards has found the often illusive sweet spot where sound biblical insight connects with wise public policy. His book *To Invent Is Divine* illuminates how, as God's image bearers, we have the privilege of being cocreators. Public policy should encourage this calling by protecting the fruit of our labors and affirming the satisfaction derived from creativity—just as our Creator saw and enjoyed His good creation. This book convincingly makes that case. Edwards holds forth a high view of the law, whose justice through ordered liberty and property leads to human flourishing while honoring our Creator. I hope readers love the ideas in this book as much as I do."

—Hon. Paul J. McNulty, president, Grove City College, and former deputy attorney general, U.S. Department of Justice

"Somehow, invention, creativity, patents, and patent law are generally considered boring and technical and of interest primarily to lawyers and technocrats. However, James R. Edwards Jr.'s new book reveals that it is a false perception. As a gifted storyteller, Edwards takes his reader on an adventure of discovery, telling the origin stories of remarkable inventions, the inventors, and how these creations have changed the world. In his chapter on Intellectual Property Basics, Edwards presents a lucid and fast-paced overview of how patents, trademarks, copyrights, and trade secret rights came to be and how they were the keys to more than 200 years of American innovation and progress. He describes this as a divine process in which ownership of one's creations is a natural right. Disturbingly, he tells us that all that is now at risk as predatory corporations and wealthy individuals have pursued predatory infringement of the intellectual property ownership of others. Edwards describes how, in a 2018 case (Oil States) before the U.S. Supreme Court, Justice Thomas wrote an opinion that effectively removed the patent ownership rights of inventors and validated the U.S. Patent and Trademark Office's use of administrative judges to remove inventors' patent protections. This is a book rich in parables. It is an unusually well-written book. Edwards is a lyrical author. I highly recommend this book for its unique insights on a vital national issue and the pleasure of reading a skilled author."

—*Pat Choate, author,* **Hot Property: The Stealing of Ideas in an Age of Globalization**

"A novel and thought-provoking meditation on the story of Creativity. In the endless diversity of mankind, Edwards identifies a universal calling to improve the world around us. For the theologically curious, certainly, this book is also a gift of understanding for inventors, tinkerers, designers, innovators, and creators—for all of us."

—Kent Lassman, president and CEO, Competitive Enterprise Institute

"Mr. Edwards's book is a refreshing take on creativity, innovation, patents, and ownership of those ideas. In addition to delving into the roots of creativity and innovation, including a case study with an artist, the book discusses an important modern aspect of innovation—those individuals working in corporate R&D on complex challenges that no single person can resolve, and the creativity, collaboration, and innovation that has been breaking new ground into sophisticated areas of science and technology."

—Susie Armstrong, senior vice president, Engineering, Qualcomm, Inc.

"Private property rights and human ingenuity—linked together and secured under law—are the primary source of progress, human flourishing, and wealth creation.

However, the current U.S. patent system is disconnecting secure private property rights from creativity, thereby disincentivizing innovation and failing American inventors."
—Hon. Thomas Massie, member, U.S. House of Representatives (R-KY)

"What distinguishes humans from all other life is our ability to reason and create. That is a natural gift imparted by our Creator on our race. With a tour de force through the history of invention and intellectual property protections, Jim Edwards draws a fascinating connection between Scripture and creativity. In the end, the creations of our mind and the fruits of our labor are uniquely human and belong naturally to, and should be owned by, the humans who create them. By methodically demonstrating that human creations are the result of natural human activities, as a result of the talents endowed by their Creator, Edwards eviscerates the concept—popular in some modern academic circles—that rights to such creations are manufactured by the State and not based on natural laws. Using inspiring examples of some of the most prolific inventors in American history, Edwards demonstrates that rights naturally belong to the inventors and the State's obligation is to protect them through the rule of law."
—Hon. Andrei Iancu, former Under Secretary of Commerce for Intellectual Property and Director of the United States Patent and Trademark Office

"Jim Edwards is a champion of inventors' rights. He's written a brilliant explanation of the biblical basis for rewarding inventions. To Invent Is Divine is a fascinating analysis of how important and spiritual our Founders' protection of patent rights is, and why we need to return to that system."
—Andy Schlafly, counsel, Phyllis Schlafly Eagles and Eagle Forum Education and Legal Defense Fund

"*To Invent Is Divine: Creativity and Ownership*," by James R. Edwards Jr., reflects a prodigious level of research into the Aha! moments of dozens of new inventions and the inventors who brought them from abstract ideas into concrete reality. "Mr. Edwards is quite obviously a devout Christian, and the book ties a number of the foundational aspects of creativity to passages and teachings found in the Judeo-Christian Bible, most notably the importance that free people be allowed to own the products of their own ingenuity. The book draws attention to the legal mechanisms of patents and copyrights originally envisioned by the framers of the Constitution for insuring such ownership, and the systematic erosion of those mechanisms in the modern economy. Given what we know to have been the significance that the Bible and Christianity played in the intellectual lives and thinking of the Founders of our Republic, Mr. Edwards's effort to define creativity in such terms is, at the very least, a provocative, stimulating, and successful undertaking. This should be equally the case for readers who may not necessarily accept all of the spiritual and mystical underpinnings of any particular religion; the discussions of creativity and imagination and the modern-day erosion of a fertile environment for their perpetuation are a story unto themselves and a most welcome contribution to our collective learning on this subject."
—Robert P. Taylor, general counsel, Alliance of U.S. Startups and Inventors for Jobs (USIJ), and owner, RPT Legal Strategies PC

"At a time when pillars of America's patent system are under assault, *To Invent Is Divine* provides the philosophical, historical, and religious basis for a course correction. This refreshing book guides us to continuing to fulfill the Founders' vision of democratic patent rights. By tapping into the creative abilities of all of our citizens and rewarding them for their vision with enforceable patent rights, we flourish together—advancing our economy, health, security, and quality of life. *To Invent Is Divine* is a timely, welcome addition to the national discussion of private intellectual property."

—*Hon. Kathleen O'Malley, Judge (ret.), U.S. Court of Appeals for the Federal Circuit*

"In *To Invent Is Divine: Creativity and Ownership*, Dr. James Edwards sets forth a rarely explored Judeo-Christian framework for understanding modern intellectual property rights systems. In patient and engaging narrative, Edwards reminds us that technological invention is a unique form of human expression—an outlet for universal creative and problem-solving impulses with which humans alone among all creatures were imbued at creation. To celebrate technological invention and artistic creativity means nothing less than to celebrate God's plan for mankind. Edwards offers a disciplined yet accessible introduction to the scriptural support for his thesis, interspersed with engaging real-life accounts of human inventors and the way their struggles and inventions, large and small, improve the human condition and contribute to human flourishing in ways that please God. Throughout, Edwards places much emphasis on explaining how human ownership of creative ideas (and the exercise of those ownership rights) is justified and shaped through Scripture and Judeo-Christian thought.

"*To Invent Is Divine* tackles a timeless idea, but it is a thoroughly modern, astute book that will appeal to a wide readership including policymakers, IP professionals, engineers, scientists, and to anyone interested in the role of Judeo-Christian values in contemporary, technology-driven society. Edwards does not just take us on a fascinating journey through the intersection of Scripture, technology, and modern law. He also describes how secular laws governing intellectual property have eroded over the past fifty years in ways that—in his words—tend to replace society's blessings of God's created order with mere fool's gold; and he offers important choices we can make individually and collectively to better reconcile today's astonishing technological advancements with a sound, coherent, and spiritual vision of the role of intellectual property rights in society. I highly recommend this important book."

—*Hans Sauer, deputy general counsel and vice president for Intellectual Property, Biotechnology Innovation Organization*

"With delightful erudition Edwards has crafted a serious work that's engaging and thought-provoking. Using elements that storytellers employ, this book introduces real-life characters like the Wright Brothers, the Gatorade inventor, and movie star Hedy Lamarr to convey an amazing history as well as a compelling argument for the biblical foundations of America's intellectual property system. A book for all creatives and those who cherish their creations."

—*Ron Maxwell, writer-director, motion picture* **Gettysburg**

"In this book, Edwards looks at patent law through a combination of theological, philosophical, and historical lenses, creating a robust argument in favor of patent rights that transcends many of today's arguments that tend to weaken intellectual property laws. Regardless of your particular views on patent law, this book has something that will be of interest and may highlight some surprising facts—I learned many new things myself."

—Kristen Jakobsen Osenga, associate dean, Academic Affairs, and Austin E. Owen research scholar and professor of law, University of Richmond School of Law

"*To Invent Is Divine* is that rarity of a promising title that exceeds expectations. Jim Edwards taps into venerable texts, ancient and modern, and especially those that shaped the founding of our nation, to illustrate how our creativity-ownership model for IP protection has made the United States the most innovative nation in history, and a technological powerhouse. In clear prose for the expert and the IP-curious alike, Jim shows that coupling creativity and ownership increases our collective store of knowledge, promotes global well-being, and enhances personal fulfillment. Jim deftly shows why James Madison was moved to say that the utility of the power to grant exclusive rights to authors and inventors would scarcely be questioned, and that the public good fully coincides with the claims of individuals."

—Brian P. O'Shaughnessy, partner and chair, IP Transactions and Licensing Group, Dinsmore & Shohl, LLP; senior vice president, Public Policy, Licensing Executives Society (USA & Canada), Inc.; chair, Board, Bayh-Dole Coalition

"The creative and innovative industries rely on strong intellectual property protections to thrive. *To Invent Is Divine* explains why and how the 'virtuous circle' of creativity and ownership functions. James Edwards thoughtfully articulates the principled foundations through which copyrights, trademarks, and patents stimulate creativity and ingenuity, and IP's vital role in a robust knowledge economy—both for creators and society. This is a book worth reading."

—Greg Saphier, senior vice president, head, Public Affairs, Motion Picture Association, and chairman, Copyright Alliance

"*To Invent Is Divine* took me on an eye-opening journey of scale regarding the idea of ownership, particularly ownership of creations and the honor and respect due to the owner/creator. At the highest scale, the author brings to life how this idea permeates God's Word and is foundational to all sorts of ethical principles laid down in the Scriptures, such as each of the Ten Commandments, and the two commandments on the honor due to parents and older people. The author connects the dots of how these principles apply in practice in a myriad of realms. For example, beside physical ownership, ownership of the purposes a creation should fulfill, ownership of ideas, and ownership of the glory commensurate with a creation—ways we rob God and each other. Or the realm of what constitutes an environment for creativity to happen—from the tiny scale of individual inventors and inventions to keeping a business running so it can do the myriad of things required to get an invention to a state of production and use where it improves human life and supports human flourishing. Not just theoretical, the book

presents historical outcomes from following or not following these ownership principles to highlight their importance and practical application. I thank God for this book and its breadth, and I highly recommend it."
—*John McCorkle, world-renowned leader and patent owner, radio-frequency (RF) engineering, and pioneer, ultrawide bandwidth RF technology*

"Jim Edwards provides readers with a compelling and fascinating case regarding the connection between human inventiveness and creativity and the divine. Invention, innovation, and creativity are responsible for the rise of our civilized world and the artistic works that inspire us. Jim's deep faith and keen understanding of the important role invention and creativity have played throughout human history make him the perfect author for this timely, must-read work."
—*Frank Cullen, executive director, Council for Innovation Promotion (C4IP)*

"As the influence of the Reformation moved through Europe, it brought the inseparable idea that individuals have God-given rights. This seed, in addition to others, begat the great spiritual and economic history of the West. Intrinsic to that unprecedented human prosperity was the notion of God's property laws. Our Judeo-Christian foundation in the West birthed a proper illumination of individual property rights. The U.S. Constitution, thanks in large part to George Washington, codified a commitment to property ownership that extended beyond land to include intellectual property. His was an indispensable role that resulted in the copyrights and patent rights unique to America. In *To Invent Is Divine*, Jim Edwards brings a necessary, timely spotlight on our divine blessing of human creativity and the fruits thereof. As the current clamor from the globalists demands, 'we will own nothing and be happy,' God's Word instructs us who has the right to benefit from or control property. 'For you shall eat the fruit of the labor of your hands, happy shall you be, and it shall be well with you' (Ps. 128:2)."
—*Diane Truitt, committee member, Ethics and Religious Liberty Committee, First Baptist Church, Dallas*

JAMES R. EDWARDS JR.

TO INVENT
IS DIVINE

*Creativity
and Ownership*

FIDELIS PUBLISHING ®

ISBN: 9781956454857
ISBN (eBook): 9781956454864

To Invent Is Divine
Creativity and Ownership

© 2025 James R. Edwards Jr.

Cover Design by Diana Lawrence
Interior Design by Lisa Parnell
Edited by Amanda Varian

All rights reserved, including the right to reproduce this book or portions thereof in any form whatsoever. For information, address info@fidelispublishing.com.

No part of this publication may be reproduced or transmitted in any form or by any means electronic or mechanical, including photocopy, recording, or any information storage and retrieval system now known or to be invented, without permission in writing from the publisher, except by a reviewer who wishes to quote brief passages in connection with a review written for inclusion in a magazine, newspaper, website, or broadcast. The web addresses referenced in this book were live and correct at the time of the book's publication but may be subject to change.

Unless otherwise noted, Scripture quotations are from (ESV) English Standard Version - The Holy Bible, English Standard Version. ESV® Text Edition: 2016. Copyright © 2001 by Crossway Bibles, a publishing ministry of Good News Publishers.

Scripture quotations marked (NIV) are from Holy Bible, New International Version®, NIV® , Copyright ©1973, 1978, 1984, 2011 by Biblica, Inc.® Used by permission. All rights reserved worldwide.

Order at www.faithfultext.com for a significant discount. Email info@fidelispublishing.com to inquire about bulk purchase discounts.

Fidelis Publishing, LLC • Winchester, VA / Nashville, TN • fidelispublishing.com

Manufactured in the United States of America

10 9 8 7 6 5 4 3 2 1

Soli Deo Gloria!

To Linda

Contents

Preface xvii

Part I: Creativity

1. A Peek into Human Creativity and Its Divine Source 3
2. Making Things and the Fruits of Your Labor 13
3. God the Creator 25
4. Made in His Image: Human Creativity 35
5. Invention at Work 47
6. Inventors and Common Grace 63
7. Creativity's By-Products: Human Flourishing 79

Part II: Ownership

8. God the Owner 97
9. Made in His Image: Human Ownership 107
10. The Mutual Reinforcement of Creativity and Ownership ... 121
11. Benefits from Owning What You Create 133
12. Property Rights 145
13. Intellectual Property Basics 157

Part III: The Patent Ecosystem

14. The American Patent System and the Iconic Inventor 169
15. The Modern U.S. Patent System 189

16. A Divide Between One's Creations
and Secure Ownership 209
17. Restoring the Biblically Based
Creativity-Ownership Link 233
18. Conclusion: Reinventing by a Biblical Return 253

Acknowledgments 267
List of Photographs and Images 269
Endnotes 271
Index 311

Preface

This started with a speech. I met the Texas leader of a public policy organization I work with at its 2019 winter leadership retreat. We talked about current issues in patent policy, and she asked me about coming to Dallas and speaking to a discipleship class at her church.

The presentation at First Baptist Church, Dallas, went well, generating a lot of interest and insightful questions. That invitation and speech launched an examination of invention and patenting—or more broadly creativity and ownership—from a biblical, Judeo-Christian perspective.

In Washington, a friend suggested I expand the speech into a book. Shortly after, I met the daughter of Gatorade's lead inventor, who brought a different perspective on these matters than those I encounter in intellectual property (IP) policy debates and discussions. This project was off to the races.

Although all this is fairly recent, my preparation for taking on this topic of creativity and ownership providentially began earlier. My interest in history, politics, and public policy, along with a desire to be guided by biblical principles in these areas, began from childhood and developed as I reached college and began my career.

In Washington to work in the Senate, various Christian ministries and circles of Christian friends made me realize how others shared the conviction of honoring the Lord through our work. My home state Sen. Strom Thurmond, Tennessee Reps. John J. Duncan Jr. and Ed Bryant provided me job opportunities in which I learned the legislative and policy process. Those jobs afforded opportunities to make wide-ranging connections, gain subject-matter knowledge—IP among others—and hone skills such as policy analysis, synthesis, and negotiation. Working on Capitol Hill and in public policy has allowed me to pursue living out my Christian faith in connection to the policy issues I've been responsible for.

Two things from early in that period were especially influential. First, Mrs. Nancy Moore Thurmond, the senator's wife, gave me a book for my birthday: Harold Lindsell's *Free Enterprise: A Judeo-Christian Defense*. Second, noted Christian intellectual Francis Schaeffer came to Washington on several occasions, where I heard him speak and met him on one occasion.

Lindsell's book models how Scripture is faithfully applied to an area of public policy. In the context of economic freedom, Lindsell discusses property rights, including intellectual property, using the whole counsel of Scripture and empirical facts. I have referred to that book over the years. Dr. Schaeffer similarly brings the whole counsel of the Bible to bear on practical problems, public issues, and broad, deep philosophical questions. Both men honor God through reason, revelation, and intellectual rigor.

Over the years, the Capitol Hill ministry Faith & Law, which I began attending in a small room in the U.S. Capitol when I moved from U.S. Senate to U.S. House staffs, has proven a marvelous forum for hearing thought leaders discuss leading issues facing the country and Christianity—speakers such as Os Guinness, Dr. R. C. Sproul, and the Rev. Richard John Neuhaus.

In the various congressional staff positions I have held, almost all involved some facet of property rights and, with several, IP. My policy and advocacy work has included IP policy, mostly patents, for a variety of clients.

While I've long sought to approach the issues I'm working on from the perspective of a Judeo-Christian worldview and the Bible's whole counsel, the invitation to do so in a speech specifically regarding patent policy caused me to focus on this familiar issue in a fresh way. The 2019 research, the preparation, the practice presentations, the speech's reception, and the degree of interest in a subject that can make eyes glaze over was encouraging and exciting.

The Schaeffers and Lindsells I've learned from have been in my thoughts as I've researched and written *To Invent Is Divine*. I've synthesized

Scripture, history, public policy, and theology to present a fact-based, biblically informed discussion of creativity and ownership, of the Divine Creator and human beings, his highest creature.

I trust readers encounter an engaging, accessible, substantive work that provides truths, insights, and facts, and that compels readers to consider things that may not have occurred to them before. And I trust all readers realize the potential benefits to the United States of restoring a patent and IP system that reflects the creativity-ownership formula found in the Bible and established by America's Founders.

James R. Edwards Jr.
Charleston, SC
August 2024

• • •

"In the beginning was the Word, and the Word was with God, and the Word was God. He [Jesus] was in the beginning with God. All things were made through him, and without him was not any thing made that was made. In him was life, and the life was the light of men. The light shines in the darkness, and the darkness has not overcome it."
— *John 1:1–5*

"He [Jesus, God's beloved Son,] is the image of the invisible God, the firstborn of all creation. For by him all things were created, in heaven and on earth, visible and invisible, whether thrones or dominions or rulers or authorities—all things were created through him and for him. And he is before all things, and in him all things hold together."
— *Colossians 1:15–17*

"In these last days he [God the Father] has spoken to us by his Son [Jesus Christ], whom he appointed the heir of all things, through whom also he created the world. He is the radiance of the glory of God and the exact imprint of his nature, and he upholds the universe by the word of his power."
—*Hebrews 1:2–3a*

"All men—lost or saved—are great in their significance. Having been made in the image of God, man is magnificent even in ruin. God made man to be responsible for his thoughts and his actions, and man fashions a significant history. This is true of both Christians and non-Christians, both men with the Bible and men without the Bible."
—*Francis A. Schaeffer, Death in the City*

• • •

Part I:
CREATIVITY

Chapter 1
A Peek into Human Creativity and Its Divine Source

"Look closely: God is reaching out to Adam from a cloud. . . . Adam is already alive, but he's listless, and he's not yet animated by the divine spark that reflects the image of God. So look closer at the cloud: . . . It's actually a brain. . . . Michelangelo was a known dissector, and this rendering is actually completely anatomically correct. So I find it fascinating that Michelangelo chose to create a link between the human mind and the divine mind."
— Phoebe Miles, referring to Michelangelo's
The Creation of Adam fresco in the Sistine Chapel[1]

"The earth is the LORD's, and everything in it; the world, and all who live in it; for he founded it upon the seas and established it upon the waters."
— Psalm 24:1–2 (NIV)

In July 1958, a new hire at Texas Instruments was so new, he hadn't accrued enough leave time to take vacation when TI's employee summer vacation period arrived. So Jack Kilby, an electrical engineer at the electronics maker, focused on solving the underlying problem then facing the electronics industry: the "tyranny of numbers."[2]

Transistors improved on the performance of vacuum tubes in electronic devices. They advanced the use of semiconductor materials in electronics and found ways to apply new basic knowledge in atomic, electrical, and materials sciences. They cut down on the size of electronic components. But each of the components of circuits had to be connected by wires for electric current to flow and make a device operate.[3] Components could be made smaller. Circuits could be designed with more complexity. But at the end of the day, there had to be interconnectivity—component by component by component. And the smaller the pieces and the more the components, the harder it became to solder all those tiny wires to make all those interconnections. Such was the vexing numbers problem.

Jack Kilby contemplated: How could you put circuit parts on a single piece of the same material and interconnect them? If it worked, this approach would end the tyrannical reign of numbers and having to solder together the interconnecting wires, which was so difficult and labor-intensive. Kilby stated the essential concept and worked out the concept's details. He made some sketches and recorded his ideas in his lab notebook on July 24. The next steps would be proving the concept by constructing a model and conducting a test.

Practically speaking, Texas Instruments made the choice of material for Kilby. The company had already outfitted its transistor manufacturing for silicon, a semiconductor material. Germanium, another semiconductor that was then the material of choice in transistors, is more stable than silicon. But germanium doesn't handle heat as well as silicon. And heat came with the territory of electrical circuitry. So Kilby fashioned a resistor from one piece of silicon and made a capacitor from another piece of silicon. This resistor-capacitor from separate pieces of the same type of material worked.

Next, Kilby created an entire electrical circuit from one silicon chip. By mid-September 1958, Kilby had built his silicon, single-chip integrated circuit. It was designed to turn direct electric current (DC) into alternating current (AC). Kilby connected the test-model silicon chip to a battery and an oscilloscope. Then Kilby—joined by his boss, the chairman of the company, and other TI executives—switched on the battery and the oscilloscope displayed a steady curvy line moving across the screen. The first all-silicon, single-chip integrated circuit worked!

A Peek into Human Creativity and Its Divine Source | 5

• • •

In 1957, Robert N. Noyce was rued as one of "the traitorous eight" by Bell Labs' savant William B. Shockley. Shockley left Bell Labs and set up shop in 1956 near Palo Alto, California. His venture, Shockley Semiconductor Laboratories, launched with a dream team of scientists and engineers who had the chops to make Shockley's invention, the double-diffusion transistor, a commercial success. He'd plucked one of his stars, Noyce, from Philco on the East Coast.

So much promise, a viable improvement to the state of the art at hand, a startup led by the co-winner of the 1956 Nobel Prize. But for all his scientific and engineering strengths, Shockley lacked talent as chief business executive. That's why Noyce and seven of his colleagues, including Gordon Moore, started looking for employment elsewhere after about a year at Shockley's shop.

Together, the eight soon jumped ship to start their own company. Bob Noyce became research-and-development (R&D) director as cofounder of Silicon Valley startup Fairchild Semiconductor. Fairchild's intended first product: a double-diffusion transistor. Noyce noodled on the tyranny of numbers and interconnectivity from time to time during the late 1950s. But first, Fairchild had to get something to market and begin a revenue stream, that is, become a going concern. A Fairchild cofounder and fellow "traitor," the physicist Jean Hoerni, soon brought Noyce an idea now known as the "planar process." Hoerni's solution would coat silicon chips with silicon oxide, which would keep them from getting contaminated—a pernicious problem with transistors at the time.

Discussions with Fairchild's patent attorney about the application for a patent on the planar process sparked broader thinking about its other possible uses. As Noyce and his colleagues thought through the planar process's potential uses, they realized the invention would make more precise interconnection possible. They saw how wiring the many connections could be supplanted by printing a conductive metal on the oxide plane instead. That would enable putting the parts of a circuit—resistor, capacitor, diode, and transistor—on a single piece of silicon—making an integrated circuit.[4]

• • •

Whoever you are in the world, chances are you own several, if not many, electronic devices. Cell phone? Laptop or desktop computer—or both? An iPad? Calculator? Television? Digital camera? Digital game system? Electric guitar or piano? Amplifier? A "smart" thermostat in your home? Or an automobile, to whose computer system mechanics connect to service your vehicle? A talking Barbie doll or stuffed toy such as Elmo? We don't think about it this way, but you own many thousands of microchips—each based on Kilby's and Noyce's solutions that changed the world.[5]

The integrated silicon chip marked an inflection point. It not only revolutionized electronics. Practically overcoming the "numbers problem" made possible things that previously were futile to implement. The microchip has made feasible so many things we take for granted: all manner of electronic devices, the modern semiconductor industry, personal computers, and much more.

This story well illustrates human creativity and ingenuity. Scholars advance basic knowledge—such as Cambridge physicist J. J. Thomson's discovering the electron in 1897. Translational work leads to identifying practical applications of a conceptual understanding—for instance, Massachusetts Institute of Technology (MIT) physics PhD William B. Shockley translating his academic understanding of electron movement into practical use as he worked with semiconductors at Bell Labs in the 1930s. Technological practitioners in industrial fields—for instance, mechanics and engineers such as Kilby—figure out how practically to put this translated information into usage and come up with something that's commercially viable. And fundamental inventions and discoveries that become the foundation of many new applications of the core technological solution—Kilby's and Noyce's cross-licensed patents for the silicon microchip, for example—act as soil nutrients, sunlight, and rain do for seeds, producing trees, bushes, flowers, and plants.

In the microchip story, this application/commercialization stage includes MIT in 1962 incorporating integrated circuitry in its Apollo Block I Guidance Computer for NASA,[6] the U.S. military's adoption in 1964 of the multifunctional integrated circuit in the complex Minuteman missile computer system,[7] cross-licensing of their patents by TI and Fairchild in 1966, Kilby's and Texas Instruments' 1967 introduction of the handheld calculator, and Noyce's and his new startup Intel's 1968 high-density memory chip.

The story also includes pioneers extending the research, invention, and application, such as Irwin Jacobs and the company he cofounded in the 1980s, Qualcomm. Cornell and MIT-trained electrical engineer Jacobs and his startup's team developed practical applications of movie actress Hedy Lamarr's code division multiple access (CDMA). CDMA enables more efficient and reliable use of radio frequencies for wireless communications, dividing spectrum in a call based on codes.[8] Actuating CDMA in telecommunications infrastructure and devices involves chipsets and other technologies. CDMA became the telecommunications industry standard, and it's been at the heart of second-generation, or 2G, and 3G wireless technologies—thanks to Lamarr, her collaborator the composer George Antheil, Irwin Jacobs, and American startup Qualcomm, which today ranks in the Fortune 500, among others.

The Source of Creativity

Why are human beings creative? What makes us capable of figuring out things, such as how to apply the properties of unseen radio waves to carry the human voice, an image, and vast quantities of data from point A to point B? Or how are we able to isolate a molecule, then use it to vaccinate people against a virus or to diagnose or treat a particular disease? Or how are we able to come up with a tune and lyrics, capture in a painting of a still-life setting what one sees or imagines, or pen an essay, or turn a play or film script into a live or film production?

Where do our creative qualities come from and how do we master them? How are we able to channel our creativity into something that fulfills the creator in us while solving a practical problem, perhaps expanding economic value and creating wealth in the process?

As the invention of the microchip shows, some discoveries hold tremendous economic worth. Some inventions spark brand-new markets for technologies not conceived of before or have only been theoretical. The question of ownership follows. Who rightfully owns an invention or discovery? Is ownership of one's own creations automatically secured? Is such ownership inherent or something determined solely by human beings? What does ownership of a new creation involve?

These questions lead to many other important questions. However, the central matter comes down to certain characteristics of God, namely his creativity and his ownership. It is no coincidence Scripture begins with a bold statement regarding God's creativity: "In the beginning, God created the heavens and the earth" (Gen. 1:1). The act of creation implies God's ownership of that which he has created. That is, creativity and ownership are interrelated. Psalm 24:1–2 states this: "The earth is the Lord's, and everything in it; the world, and all who live in it; for he founded it upon the seas and established it upon the waters" (NIV). This interrelation of creativity and ownership is true for God, but also for human beings.

With regard to creation, some things have intrinsic value, while others are more common or mundane, only deriving value extrinsically. Some creations are more pleasing to behold, while others are plain or even unsightly. Some created things are more useful than others. Some creations have little apparent redeeming value or use. Other things are outright ugly, dangerous, or repulsive. This is certainly so after the Fall, when the first man chose to disobey God's direction and thereby disrupt the "shalom" peace originally covering all aspects of life and the creation.[9] Nevertheless, at its core, creativity has positive qualities; it is productive, constructive, perhaps ennobling, pleasing, and of benefit.

Truth, beauty, goodness. Creating expresses truth. There is beauty in creating something, and typically in a creation. There is goodness in creating. The Lord sets the premier example of this. The creation account in Genesis 1 tells how God created the heavens and the earth. Each day recounted tells what work of creation God performed at that stage. On the first day, when the Lord created light, he "saw that the light was good." When he created the dry ground and the seas, God observed that this, too, was good. He called the vegetation he created good. His creations in the heavens, the sun, moon, and stars, God saw as good. Also deemed good were the fish and the birds he made to fill the waters and the skies. The land animals and moving creatures he made he called good too.

The Lord's capstone creature, human beings, he made in his own image—an indication of special dignity and honor. "So God created man in his own image, in the image of God he created him; male and female he created them" (Gen. 1:27 NIV). The next chapter of Genesis elaborates on his creation of human beings. It tells how the Lord "formed the man from the dust," breathed life into him, and similarly fashioned the woman from a rib taken out of Adam. Adam's reaction to seeing Eve boils down to a well-considered "Wow!" Having begun working the Garden of Eden (work is good; Gen. 2:15 gives God's initial charge of human effort for productive purposes) and having assigned names to all the animals, Adam understood God's point that "it is not good for the man to be alone," lacking human companionship. He would now appreciate God's custom-designed companion, with God making "a helper suitable for him" (Gen. 2:18 NIV).

The first man's informed response to God's display of creativity done expressly for him was to say of the woman, "This is now bone of my bones and flesh of my flesh" (2:23). She was incomparable; none of the birds, fish, livestock, or beasts came close. This perfect act of creativity for his image-bearer creature represents the essence of ingenuity. Eve, as a human being and as a creature, filled a void Adam hadn't realized existed. God dramatically demonstrated this need, this void, this problem. He then created, producing a unique, perfect complement to the male human model. She met the need, filled the void, solved the problem. The female model of human form displayed God's intended solution for Adam, she also bearing God's image. There was truth, beauty, and goodness in her likeness, similar to, but also unlike, those divine characteristics as manifested in the man.

To be sure, human beings mark the pinnacle of God's created order. However, lower-order creations also reflect the imagination and skill of the Creator through his handiwork. The heavens, the earth, the fauna, and the flora are in some measure invested with truth, beauty, and goodness. The psalmist David captures in Psalm 8 the majesty, the reflection of God's glory, the sublime qualities of the creation. It reads, in part: "When I look at your heavens, the work of your fingers, the moon and the stars, which you have set in place, what is man that you are mindful of him, and the son of man that you care for him?" (vv. 3–4). The vastness and majesty of his creation that we observe in nature, in the night sky, in the nooks and crannies around us, in the viewfinder of the most powerful microscope or of the most powerful telescope—these can cause human beings to feel minute, humble, even unimportant. The Creator has "set [his] glory above the heavens" (v. 1), adding to our sense of insignificance in the scheme of things. Yet, the psalmist reminds us of mankind's place in the creation order, echoing Genesis: "You made [humans] rulers over the works of your hands; you put everything under their feet: all flocks and herds, and the animals of the wild, the birds of the air, and the fish of the sea, all that swim the paths of the seas" (vv. 6–8 NIV).

This attests to the different purposes of each element of God's creation. It also illustrates the Lord's endowment of certain qualities in human beings. The infinite Creator made finite human beings in his image. Creativity counts among his divine attributes. Thus, creativity is divine, one of the divine qualities endowed in each human being. We all have creativity. Though finite, people are given the ability to create and thus the right to own property, starting with the things one creates.

When God completed the works of creation, "God saw all that he had made, and it was very good" (Gen. 1:31 NIV). Very good. God's creation as a whole and its constituent parts embodied truth, beauty, and goodness. This is the model for creativity. And it is reflected in the creativity and ingenuity human beings possess.

● ● ●

The premise of this book is that human creativity reflects the creative characteristic of God the Creator, who has endowed his highest creature with this quality and placed within human beings an inherent awareness of individual ownership. Indeed, human beings possess inborn rights of ownership of their own property, including those things one creates or lawfully comes to own. These central, shared characteristics—creativity and ownership—lead to answers to the questions that naturally follow, some of which are stated above.

Moreover, these human characteristics, human limitations—we are finite creatures made by and in the image of an infinite Creator—and other constraining factors have led to the discovery and development of concepts and principles such as property rights. When human laws, institutions, and practices relating to creativity and ownership more closely align with divine principles, benefits follow. Individual creators can explore and discover and make secure in the knowledge that what they create will belong to them. Thus, by understanding the foundational biblical principles of

God's creativity and ownership of his creation, human creative endeavors can more fully avail both the creators and society.

The topics of God's creativity and ownership—those same endowments uniquely given to the creatures bearing his image, as well as how creativity and ownership come together as intellectual property—are discussed in depth in subsequent chapters.

• • •

If you're interested in innovation, particularly in deepening your understanding of the biblical basis for human creativity and ingenuity and how we secure ownership rights to the fruits of our labor, then you've found the right book. King Solomon's rather depressing piece of wisdom literature, the book of Ecclesiastes, states, "There is nothing new under the sun" (1:9). However, much of what is under the sun may be patentable (famously stated as "anything under the sun that is made by man")[10] as new or improved and useful. More broadly, creativity in the useful and the fine arts offers a lot of stuff to work with. The focus here remains primarily on creativity and ownership related to invention and patents.

This book contains three parts. The first focuses on creativity, the second on ownership, and the third on the patent ecosystem. Part I examines God's creativity and the creative spark within human beings. Topics include the Creation Mandate, the invention process, and Common Grace. Part II discusses God's ownership of what he has created and inherent rights of ownership humans possess, along with how creativity and ownership mutually reinforce each other to promote human flourishing. Then ownership in the context of property rights and intellectual property is introduced. Part III considers the patent ecosystem. We examine the American patent system, its "golden age," the diminution of "secure" exclusive rights, and restoration of the biblically rooted linkage of creativity and ownership.

Thus, we hold up to the light a multifaceted object. We turn it like a prism to see more completely creator and creature, creativity and ownership. These facets are illuminated in different light, in different colors. Examples of inventors and inventions throughout the book elucidate facets of our exploration and bring the topics to life.

Chapter 2
Making Things and the Fruits of Your Labor

"You shall eat the fruit of the labor of your hands."
— Psalm 128:2a

Human creativity manifests itself across time and places. Human creative expression is found in virtually every society and every epoch, from ancient cave-dwellers to tribes ensconced in deep jungles, from Mesopotamian civilizations before Christ to Sinhalese and Asian societies to their contemporaries in the Greek and Roman empires, from the Renaissance in Europe to the arts and crafts of plains Indians in what would become the United States to the heights of Western civilization from the Protestant Reformation through the twentieth century.

Human creativity manifests itself in many ways. It encompasses visual arts such as painting, drawing, photography, motion pictures, television productions, and sculpture. It includes music, dance, poetry, literature, textiles, and jewelry. It ranges from folk tunes passed along from generation to generation in a clan or society to Handel's *Messiah* and *Water Music*, Vivaldi's *Four Seasons*, Bach's concertos, oratorios, toccatas, and fugues, Mozart's *Eine Kleine Nachtmusik*, from child prodigy Alma Deutscher's inspired concertos, operas, and chamber pieces to Irving Berlin's popular patriotic songs and shows (his first song published at age nineteen and he composed into his seventies).[1] It involves simple, homemade instruments and well-crafted Stradivarius violins, Martin guitars, and Steinway pianos. Human creations come in black-and-white and color, one-, two-, and three-dimensional, acoustic and electric, live and recorded, analog and digital.

Human creativity combines time, energy, effort, thought, financial and other resources, analysis, synthesis, insight, and inspiration. Michelangelo, when in his twenties, spent a year sculpting the *Pieta*, three years working on his *David*, then four years painting the fresco on the Sistine Chapel's ceiling, each work on commission.[2] Leonardo da Vinci invested his talents and abilities in many productive endeavors, including in art and invention.[3] Da Vinci painted *The Last Supper* and the *Mona Lisa*, among other artistic works, as well as designing flying machines, an adding machine, and other devices.

Handel composed *Messiah* in its entirety in an estimated twenty-three days—an exceptionally quick pace.[4] William Shakespeare, the Bard of Avon "not of an age, but for all time," authored some thirty-nine plays, one hundred fifty-four sonnets, and other works. Shakespeare's works span comedies, tragedies, histories, and tragicomedies. Many count among the world's best-known stories, such as *Romeo and Juliet*, *Julius Caesar*, *King Lear*, and *A Midsummer Night's Dream*.[5]

The Wright Brothers built a wind tunnel and tested the performance of various wing shape, curvature, length, and width. This gave them an empirical edge in designing their next flying machines, including the 1903 milestone Flyer. It was "the first in the history of the world in which a machine carrying a man had raised itself by its own power into the air in full flight, had sailed forward without reduction of speed, and had finally landed at a point as high as that from which it started."[6]

* * *

Application of the laws of nature through the scientific method brings us to discovery and invention—the practical side of science in the realm of the objective universe. Invention centers on solving specific problems. It has to do with working out the "new" and "improved." The language of the U.S. Constitution, regarding the work creators' advancing "science" (in the broader sense of knowledge) and of inventors' discoveries, refers to the latter as advancing "useful arts."[7] This category (e.g., mechanics, materials science, electrical engineering, biotechnology, computer sciences) contrasts somewhat with the fine arts, where works and tastes tend to the more subjective.

Depending on the specific useful art, invention and discovery display ingenuity through the applied use of mathematics, science, technology, and engineering. Useful arts, discovery, and invention relate to observing natural phenomena, daily situations, rudimentary implements, and practical problems, and applying knowledge and thought to solving specific, identified problems. Thus, invention takes center stage in this chapter's overview of human ingenuity. Here we briefly discuss key concepts, putting them in context and relationship with one another. Later chapters delve into greater depth on the fundamental topics.

The Process of Invention

The process of invention in virtually all areas follows roughly the same path. Invention doesn't look the same as with a cookie cutter, follow rigid steps, or involve exactly the same linear pieces. However, invention tends to have certain aspects in common. An informative paper from the Lemelson Center for the Study of Invention and Innovation about the process of invention says "invention takes place over time and is the result of the inventor doing things—thinking, building models, sketching, and conducting experiments."[8] Whatever shape a specific inventive effort may take, applying one's intellect, knowledge, and skills in a creative process leads to discovery or development of an idea to pursue.

Invention often starts with a problem. An inventor identifies an unmet need or an unsolved problem, or perhaps arrives at an insight from his or her work or study. Inventors make use of scientific or technical knowledge, which they possess from their education, training, and experience or gain from keeping abreast of the technological literature, in coming up with practical solutions. They have an idea for solving the problem. The idea may flow from systematically pursuing the scientific process, as a discovery or by-product of benchwork, from accident or serendipity, from conversations with colleagues or advisors, from iteration, from an epiphany in which a realization or envisioned potential solution springs forth, or from imagination.

Invention progresses by carefully noting ideas and observations, drawing a sketch, making a prototype, testing and revising it, and taking detailed notes on each hypothesis, learning, discovery, or iteration. Jack Kilby proved the concept of making a resistor-capacitor from separate pieces of silicon in the Texas Instruments lab. Trial and error characterize invention and discovery. Models, materials, tweaks—each of these and more are brought to bear in the invention process. Inventors "mingle ideas and objects." Inventors frequently employ "a variety of substitutes—rather than make a full-scale version, an inventor might use a prototype, sketch, series of calculations, computer simulation, or even a written description."9

Inventors may discuss their ideas and aspects of an invention along the way with others. These confidants may be a research assistant, a business partner, an intellectual property attorney, a prospective investor, or a trusted advisor. However, inventors must be discreet. A public disclosure renders their discovery unpatentable. They often swear those with whom they discuss their invention to secrecy with a nondisclosure agreement. Each phase of inventing and private discussions influence new pivots, new tests, new insights, new directions toward a promising, tangible new or improved device, material, process, or discovery. Once an inventor develops a viable invention, he or she typically applies for a patent. Like a deed, a patent secures the newly created property within the boundary lines of the invention as his or her exclusive property.

With clear title secured, commercialization proceeds. Commercialization involves the work necessary for bringing the invention to market. As the popular television program *Shark Tank* shows, some inventors start a business around their invention, bootstrapping their enterprise or raising investment funding. Others look for a manufacturer to whom to license their patent, make the product, and develop a commercial market for it. Others sell their patent to an established commercial player or another entity.

Commercialization is where the invention moves from new, useful solution through product development and market development to ultimately something consumers or industrial entities accept and adopt. It amounts to diffusion of a new product in the commercial market. The earnings from patent licensing fees, royalties, etc. typically repay investors at a return, produce revenues for an entrepreneurial inventor's startup, and fund ongoing research and development (R&D) into improvements of the original product, process, or material or the continued invention of varieties of or accessories for the original product.

Similar to the process of invention, the scientific method involves a systematic approach to gaining new knowledge by developing and testing hypotheses. There is observation. Inductive reasoning, involving a premise from observations, leads to a research question. A hypothesis follows, proposing a testable relationship among certain variables. An experiment tests the validity of the hypothesis. Analysis often employs statistical calculations to ascertain the strength of the variables' relationships and the significance of the hypothesis. The results of an experiment provide empirical evidence of the actual relationship among the variables and the effects of independent variables on the dependent variable. The analytical results of the experiment, which may be expressed statistically or as a deduced premise, lead to conclusions or findings. This process typically takes place before invention.

Such inquiry is more or less for the sake of advancing knowledge. The findings are shared with others, such as in a research paper or a peer-reviewed journal article. This use of the scientific method, occurring mainly in academic settings, is known as basic research. It extends human understanding of phenomena, such as the correlation of a certain molecule with a certain disease.

Findings in basic research may lead to practical application. For example, Dr. Sid Malawer studied under University of Florida professors of medicine, including Dr. Robert Cade, a specialist in renal medicine, in the 1960s. Malawer's research examined how sodium moves through the body. Malawer discovered that two sodium molecules are ushered into the body via a glucose molecule that docks in the epithelial cells of the intestine, which opens a type of "cellular elevator" or channel. Once the sodium molecules enter, they hold up to 300 molecules of water in the intestine. This finding advanced knowledge in this area.[10] It also contributed to a commercially valuable application, informing the development of the Gatorade (named for the UF mascot, the Gators) formulation for restoring water and electrolytes lost by athletes.[11]

In any event, invention "is trying to merge abstract ideas in the mind with objects in the real world." This involves "models, sketches, and notes as well as machines . . . in various stages of development."[12]

Another example of a common means of discovery and invention comes from observing nature. Human flight was inspired by watching birds. Da Vinci sketched drawings of flying machines with movable wings. One of the aeronautic leaders of the late nineteenth century, Otto Lilienthal of Germany, designed his flying machine with movable—i.e., flapping—wings. Flapping wings didn't prove feasible for human flight. Lilienthal eventually switched to hang gliders with fixed wings.[13] Rather, applying the learnings of aeronautical inventor Sir George Cayley and others, fixed-wing design stemmed from his and others' efforts with gliders.[14] Cayley studied birds in flight, eventually focusing on their gliding rather than their flapping of wings. By 1804 Cayley was pioneering gliders. The Wright Brothers opted for fixed-wing designs, adapting insights from the laws of nature in God's creation. The Wrights had both mechanical ability and intelligence, it was observed, which served them well as aeronautical inventors.[15]

• • •

At heart, inventors, themselves created by an infinite, personal God, apply their inherent creative qualities with observation, reason, and thought in pursuit of creating new or improved solutions to certain problems. Some inventions are incremental, others complementary, and still others monumental. Some adapt what has been observed in nature. Inventors apply the knowable laws of nature in one way or another, employing their knowledge and skills in practical problem-solving.

Whether they realize it or not, those endowed with such scientific or mechanical abilities, employing the talents the Creator has given them, are doing what they were designed to do, reflecting the creative attributes of God and endeavoring in a corner of his creation. In fact, their efforts of ingenuity comport with the Creation Mandate God gave Adam and all humankind. This combination may lead to inventions, discoveries, and breakthroughs of something that didn't exist before. Their creativity results in the creation of new property—the fruits of their labor.

Owning the Fruits of Your Intellectual Labors

Human creativity and productivity, as well as private ownership rights, are part of God's "Common Grace." Among other things, Common Grace means the laws of science and mathematics apply generally for practical purposes; it isn't special knowledge reserved for people who hold Judeo-Christian beliefs; and the benefits are enjoyed generally. Some atheists and agnostics have scientific or mechanical aptitude, as do some Christians. Other people, some believing in Jesus Christ and some not, may lack aptitude in these areas while being talented in literature, music, business management, or marketing, for example.

Human beings are designed and called to be productive. God's Creation Mandate (Gen. 1:28 and 2:15) charges humans not only to survive, but to thrive, to create abundance. And individuals have a stake in what their efforts yield. Isaiah 65:21 says, "They will build houses and dwell in them; they will plant vineyards and eat their fruit" (NIV). This and similar Bible passages imply a person's enjoyment or gain from what that individual puts into something—the output, a dwelling, fruit.

This is apart from an individual's standing with God. Effort expended working on creative projects results in something new. Labors produce fruit. This happened with inventing the microchip for Kilby and Noyce in how they each enhanced and combined semiconductive materials for superior interoperation. The human effort, applying one's intellect, insight, knowledge, reason, skills, etc., adds value to raw materials. Andy Crouch captures the adding of value, "the transition from nature to culture as a move from 'good' to 'very good.' Eggs are good, but omelets are very good. Wheat is good, but bread is very good. Grapes are good, but wine is very good."[16] Inventors, creators, and discoverers produce fruit from their labor, add value by their creative results, expand human knowledge—and rightly have ownership claims.

Among the fruits of one's labor, including intellectual labor, are joy and a sense of accomplishment, while ownership delivers tangible fruit. We get fulfillment from investing ourselves in honest, productive work. There also comes joy from the insights gained, sometimes in profound ways.

The expression "the fruits of one's labor" recalls a farmer's efforts planting, watering, and aerating soil, yielding a crop that belongs to the person who owns the land and works it to produce a harvest. Founding Father John Dickinson, referring to a scriptural metaphor, put it, "the foundation of all [private property rights is] that their property, acquired with so much pain and hazard, should be disposed of by none but themselves—or, to use the beautiful and emphatic language of the sacred [S]criptures, 'that they should sit *every man* under his vine, and under his fig tree, and *none should make them afraid*'" (emphasis added).[17] The work represents adding value to raw or unfinished material, with grapes and figs the fruit.

An inventor or creator labors to make something that either didn't exist before or that is improved. This manner of creation of wealth by human effort rightly and justly benefits the creator, just as physical labor benefits a worker and the person or persons who trade the fruits of their own labor for it. The "fruits" from one's inventive labors constitute property not previously in existence. They are intellectual fruits. A human creator's ownership of what he or she creates tracks with God's ultimate ownership over his creation as previously established. God created the universe. All of it belongs to him. Generally, ownership rights are inherent to the person who creates new property.

Landowners, automobile owners, those owning money or other belongings each possess rights of ownership over their property. To secure these property rights, they need clear title. Such documents as deeds, titles, and patents describe the property, its boundaries, perhaps other features or characteristics. Without clear title, one can't build on or sell "real" property, dispose of financial instruments, or enforce rights against those trespassing on or appropriating one's property.

When we honor and respect private ownership, this supercharges creativity and its derivative societal and economic benefits. Secure private property rights democratize and incentivize invention and creativity. This in turn leads to new and better "fruits." In many instances, these benefit others—through economic exchanges, sharing, or collaborations achieving more than the creator alone could accomplish. The democratization of property rights, secured through legal protections and deeds held by the creator or subsequent buyer of the property, promotes equality under the law. This has enabled American inventors who were black or women to secure title to their inventions and creative works through patents and copyrights.

Democratized ownership rights spur untold numbers of people who have an original idea and develop it into something tangible to pursue what they might otherwise not have bothered with, had governing authorities been able to expropriate the fruits of their labor. Christian writer and editor Harold Lindsell explains the socialistic opposite of democratized private property rights: "In the broadest sense socialism is opposed to private ownership of the means of production. . . . Marxism and utopian socialism share the same ultimate objective, i.e., the organization of society in such a way that the workers own, control, and operate everything cooperatively."[18] Lindsell observes that communism and socialism never work out to be egalitarian or cooperative in practice. The real-world result of centralized government and lack of freedom and of secure private property rights is captured in the Soviet quip: "They pretend to pay us and we pretend to work."[19] Socialism robs citizens of the incentive to work, be creative, and be productive. And both individuals and the entire nation suffer the consequences.

Foreign governments' expropriation of private owners' intellectual property usually leaves those countries without much of an intellectual property–centered industrial base or much domestic innovation. The United States has seen similar harmful effects. By 1980, the U.S. government had plowed billions of dollars into research. This resulted in tremendous expansion of knowledge. Uncle Sam owned 28,000 patents, but less than 5 percent of those inventions were even attempted to be put to practical benefit. Private companies couldn't afford the risk of licensing these patents without reliable, exclusive property rights to the fruits of their labor. The Bayh-Dole Act fixed this. It secured university and private researchers' ownership and licensees' exclusivity of discoveries and inventions that stemmed from federal research funding.[20]

To secure the rights to the fruits of one's intellectual labors, one must prove the new creation is his or her intellectual property (IP). IP, typically in copyrights, patents, or trademarks, serves as title or deed to specific property—in these cases a work such as a musical composition, a sound recording, a film production, or a book manuscript; an invention, a discovery such as an isolated molecule, or a technological improvement;

or a word, a name, or a symbol that represents a company or its goods, respectively. Another form of IP, trade secrets, is valuable, closely guarded, confidential business information. Contracts protect trade secrets from becoming publicly known.

In contrast to trade secrecy, patents, copyrights, and trademarks balance exclusive property rights with public disclosure. The patent system is the original open-source model. In exchange for exclusivity to the discovery or invention for a limited period of years, the patented discovery's or invention's conceptual and technical details are publicly disclosed so others may immediately learn from the invention. Other scientists and engineers skilled in the art may freely invent around or improve upon the invention described and taught in a patent. The claims specified in the patent detail the boundaries of the newly created property. And when patent and copyright terms expire, the works enter the public domain.

The "Patent Bargain" of exclusivity in exchange for disclosure achieves the intention the Founders stated in the Constitution's IP Clause: "To promote the progress of science and useful arts . . ."[21] This private property rights-based exchange takes the long view. It preserves, protects, and promotes both creativity and ownership for the ongoing, derivative benefit of society. And it fits with the biblical principles of both God's and human beings' creativity and ownership.

The U.S. patent system, as the Founders designed it, rests upon private property rights. Its democratized IP protections, secured for the "first and true inventor" or creator to one's original works, underscores this principle. The U.S. patent system, by securing exclusive and enforceable rights, is biblically aligned because it honors individuals' creative qualities, which God bestowed upon us. The American patent system respects humanity's divine gifts, as manifested in individual creativity and ownership, by issuing clear title to the tangible fruits of one's creative endeavors.

This marriage of creativity and ownership, most notably as achieved in the United States, yields countless offspring that benefits humanity. Creativity without private ownership nor ownership without creativity, throughout history, has sufficiently incentivized the level of human achievement found in America. Abraham Lincoln, the only president with a patented invention, eloquently captures the key to the success of this fruitful marriage: "The [U.S.] patent system . . . secured to the inventor, for a limited time, the exclusive use of his invention; and thereby *added the fuel of interest to the fire of genius*, in the discovery and production of new and useful things" (emphasis added).[22] Talk about a marriage made in Heaven!

* * *

This combination of private ownership and individual creativity joins divine endowments in individual human beings. This brings to mind the Declaration of Independence's Preamble where it refers to "inalienable rights" that include "the pursuit of happiness," widely understood as private property ownership. The offspring of individual creativity and property rights brings the joy of creativity and the joy of owning one's own creations. This divine gift—the combination of creativity and ownership—results in human flourishing, individually and societally.

Chapter 3
God the Creator

"[God] has made everything beautiful in its time.
He has also set eternity in the human heart; yet no one
can fathom what God has done from beginning to end."
— Ecclesiastes 3:11 (NIV)

"There is nothing new under the sun," Ecclesiastes 1:9 says. Then where did everything under the sun originate? Where did the sun come from? And the moon, the stars, the planets, the galaxies, the universe? From something? From nothing?

What do we know or have we considered about God's creativity? How could that provide insights into the creativity of human beings? Think about watching a brilliant sunset over a lake, a mountain, or the ocean. It's like a vivid, dynamic art display in real time—except it's unique, steadily changing, and once finished, gone forever.

Similarly, think of a blooming flower, how the hues of its petals may appear uniform from a distance, but closer inspection reveals interesting features, patterns, blends that too are unique and, once withered, are gone. Such examples of the Creator God's creativity can help us understand human creativity more fully.

Divine creativity is foundational because creativity is a characteristic of the God of the Bible. Considering God's creativity gives us an inkling of the Creator who is infinite and unlimited. These characteristics of God, revealed in Scripture and in nature, inform our understanding of God's ability to make things out of nothing—something humans cannot do.

We begin with the bookends of the Bible, Genesis and Revelation. We'll consider portions of several Christian confessions and creeds, then look at biblical references to God's creativity in the Old and New Testaments.

The God Who Creates Something from Nothing

The Genesis account of creation begins with a statement of the eternal, infinite God creating the heavens and the earth. The first verse of the first chapter of the first book of the Bible says, "In the beginning, God created the heavens and the earth." It continues, "And God said, 'Let there be light,' and there was light. And God saw that the light was good. And God separated the light from the darkness. God called the light Day, and the darkness he called Night" (Gen. 1:3–5a).

God went on to create the seas and waters; vegetation; the sun, moon, and stars; fish and birds; all the creatures on earth such as livestock and beasts; and, in his own image, God created male and female human beings. In a summation of his creation of human beings, Genesis 1:27 says, "So God created man in his own image, in the image of God he created him; male and female he created them."

This account indicates an important difference between God and human beings, as well as the unbounded nature of God's creative ability in contrast to humankind's limitations, creative and otherwise. God is infinite. The Westminster Confession of Faith says, "There is but one only living and true God, who is infinite in being and perfection, . . . eternal, incomprehensible, almighty, most wise, most holy, most free, most absolute, working all things according to the counsel of his own immutable and most righteous will" (2:1).

Being infinite comes with benefits. Orthodox (meaning Bible-believing, originalist) Christians generally understand the infinite God created the universe and he made it out of nothing. The triune God spoke his creation into being, *ex nihilo*, from nothing. Without the divine attribute of infinitude, creation from nothing would be unachievable. The seventeenth- and eighteenth-century Protestant theologian and author of one of the most highly regarded commentaries on the Bible, Matthew Henry, explains:

The manner in which this work [in Gen. 1:1–2] was effected; God *created*, that is, made it out of nothing; there was not any pre-existent matter out of which the world was produced. The fish and fowl were indeed produced out of the waters, and the beasts and man out of the earth; but that earth and those waters were made out of nothing. By the ordinary power of nature, it is impossible that something should be made out of nothing; no artificer can work unless he has something to work on. But, by the almighty power of God, it is not only possible that something should be made out of nothing, (the God of nature is not subject to the laws of nature,) but in the creation, it is impossible it should be otherwise; for nothing is more injurious to the honor of the Eternal Mind than the supposition of eternal matter. Thus the excellency of the power is *of* God, and all the glory is *to* him (emphasis in original).[1]

Scripture passages such as Psalm 33:6, 8–9 confirm creation ex nihilo: "By the word of the Lord the heavens were made, and by the breath of his mouth all their host. . . . Let all the earth fear the Lord; let all the inhabitants of the world stand in awe of him! For he spoke, and it came to be; he commanded, and it stood firm." The ex nihilo theme also appears in the book of Hebrews: "By faith we understand that the universe was created by the word of God, so that what is seen was not made out of things that are visible" (11:3).

The first bookend, Genesis 1, is the foundation for understanding God as the Creator. It gives the first indication of the ex nihilo nature of this divine creativity. "And God said. . . . And it was so." Genesis 2 notes the foregoing creations represent God's creative work. God is the originator, the ultimate Author, the Maker.

This also shows from where human beings come to take pleasure in their creativity. The Creator takes pleasure in his own creative efforts. It appears throughout the first chapter of Genesis. The Creator God observes his handiwork at each stage, declaring it "good." In the second chapter, after the sixth day, "the heavens and the earth were finished" (2:1). Upon completion of this project, God takes it all in—in a sense, enjoys the fruits of his labor. God calls it "very good" (1:31). This reflects how creating is a fulfilling endeavor, in common between God and man. It is the basis of human flourishing.

The final book of the Bible, Revelation, recounts praise for God's creativity. Heavenly worshippers around God's throne say, "Worthy are you, our Lord and God, to receive glory and honor and power, *for you created all things, and by your will they existed and were created*" (4:11, emphasis added).

After the Judgment and the unveiling of the new heaven and the new earth, God reveals "the holy city, new Jerusalem" (Rev. 21:2). This new and improved Jerusalem is the magnificent creation God makes and presents at the dawn of eternity as the temporal passes away. Even at that point, the Lord announces from his heavenly throne, "Behold, I am making all things new" (21:5). Note the present progressive verb tense, "am making," in the English Standard, New International, New American Standard, and other versions. Might that mean the Creator God is not completely finished with his creative endeavors? Scripture does not say that, but it's a reminder that he is the ultimate Creator.

Thus, Scripture begins and ends by continual references to God as Creator; creativity clearly characterizes the God of the Bible. Even after the "big reveal" of the new heaven and earth following Judgment, he says, "Behold, I am making . . ." How intriguing.

• • •

This scriptural basis for knowing God is the Creator of the universe and its contents, including human beings, has been affirmed and encapsulated throughout Christian history. A quick survey of early church creeds and confessions,[2] which are based on Scripture, reiterates this understanding of God's creativity. They uniformly refer to the Heavenly Father as "maker of heaven and earth" or that essential concept. These faithful works indicate not only God's creative quality, but also his infinity, his majesty, and his ability to make tangible and intangible creations from nothing. This combination of characteristics recurs in these affirmations.

For instance, the Apostles' Creed begins, "I believe in God the Father, Almighty, Maker of heaven and earth." Similarly, the Nicene Creed starts, "We believe in one God, the Father Almighty, maker of heaven and earth, of all things visible and invisible."

The Westminster Confession of Faith addresses the creation: "It pleased God the Father, Son, and Holy Ghost, for the manifestation of the glory of his eternal power, wisdom, and goodness, in the beginning, to create or make of nothing the world, and all things therein, whether visible or invisible . . ." (4:1).

The Heidelberg Catechism's answer to question twenty-six about God the Father reads: "That the eternal Father of our Lord Jesus Christ (who of nothing made heaven and earth, with all that is in them; who likewise upholds and governs the same by his eternal counsel and providence) . . ."

The Belgic Confession, Article 12, says of creation: "We believe that the Father, by the Word, that is, by His Son, created of nothing the heaven, the earth, and all creatures as it seemed good unto Him, giving unto every creature its being, shape, form, and several offices to serve its Creator; that He doth also still uphold and govern them by His eternal providence and infinite power for the service of mankind, to the end that man may serve his God."

These confessions and creeds, drawn from Scripture's teachings, aver that the Creator God made the heavens and the earth and its contents ex nihilo. To bring into being something a human inventor conceptualizes is remarkable. To create the universe in all its multitude of detail, as God has, is beyond human comprehension. To begin to take in the fact that God made all these things stretching as wide as the universe, some as tiny as submicroscopic and subatomic particles, quickly becomes an exercise in humility. It should inspire awe of the Creator.

The takeaway is the infinite, eternal, personal God made the earth, the heavens, animals, plants, and people. Moreover, God created the heavens and the earth out of nothing. Knowing this orthodox Christian belief will serve us well as we turn to the Creator God's creativity, as further described in Scripture and seen in creation.

• • •

As seen, God is creative. From Genesis to Revelation, Scripture registers this fact emphatically. We've observed it begins in Genesis chapters 1 and 2 with the creation story. "In the beginning, God created the heavens and the earth" (Gen. 1:1).

Scripture's opening and closing books underscore God's being a Creator, one unbounded, unlike his limited creatures. These bookends invite a look throughout the Bible to discover more about God's creative nature. Considering the breadth of such references, the variety of contexts, and the ways in which the Creator and his works are described may deepen our appreciation for what being creative means for human beings.

The Divine Creator Between the Bookends

A variety of biblical passages from the middle sixty-four books acknowledges the sovereign God as Creator of the heavens and the earth. In some, God speaks of his creativity. In others, a human being remarks about God's creativity and creation, and credits him as the maker. No matter their context or the speaker, these examples of passages about the divine Creator enrich human understanding of who this Creator God is. These Scriptures between the bookends of Genesis and Revelation illuminate infinite creativity. Moreover, the richness and variety of these excerpts give us a window on the Person characterized by infinity and unlimited creative ability. We begin with examples from the Old Testament.

The prophet Isaiah delivers God's bold, clear message that he created the heavens and the earth without any help from anyone else. "Thus says the Lord: 'Heaven is my throne, and the earth is my footstool; what is the house that you would build for me, and what is the place of my rest? *All these things my hand has made, and so all these things came to be*, declares the Lord'" (66:1–2a, emphasis added).

In Isaiah 40:25–26, God cites his creation of the stars to illustrate his authoring and governing the heavens. "To whom then will you compare me, that I should be like him? says the Holy One. Lift up your eyes on high and see: who created these? He who brings out their host by number, calling them all by name; by the greatness of his might and because he is strong in power, not one is missing." The message is clear: God alone created the heavens and needs no human help running his universe.

Again, the prophet Isaiah conveys the Lord's statement of being the Creator and ruler of his creation: "I form light and create darkness; I make well-being and create calamity, I am the Lord, who does all these things. Shower, O heavens, from above, and let the clouds rain down righteousness; let the earth open, that salvation and righteousness may bear fruit; let the earth cause them both to sprout; I the Lord have created it" (45:7–8).

Isaiah intertwines God's creative acts in time and space to convey the validity of the Lord's word about himself. "For thus says *the Lord, who created the heavens* (he is God!), *who formed the earth and made it* (he established it; *he did not create it empty*, he formed it to be inhabited!): 'I am the Lord, and there is no other'" (45:18, emphasis added).

Isaiah 48:12–13 is another instance where God points to his visible creation to prove the fact of his divine sovereignty. "Listen to me, O Jacob, and Israel, whom I called! I am he; I am the first, and I am the last. *My hand laid the foundation of the earth, and my right hand spread out the heavens*; when I call to them, they stand forth together" (emphasis added).

In Isaiah 40:28, the prophet makes the same divine Creativity his evidence of God's incomparable majesty. "Have you not known? Have you not heard? The Lord is the everlasting God, *the Creator of the ends of the eart*h. He does not faint or grow weary; his understanding is unsearchable" (emphasis added).

Scripture between the bookends also highlights God's creation of human beings and his regard for his highest creature. For example, "Have we not all one Father? Has not one God created us?" (Mal. 2:10a). Of how he will provide for his people such as with water, the Lord expresses why: "that they may see and know, may consider and understand together, that the hand of the Lord has done this, the Holy One of Israel has created it" (Isa. 41:20).

More specific aspects of God's creative works, the organs of human speech, other senses, and, implicitly, of human rationality, are cited to assure Moses of God's promise to go with Moses on his mission in Egypt. Exodus 4:10–12 says, "But Moses said to the Lord, 'Oh, my Lord, I am not eloquent, either in the past or since you have spoken to your servant, but I am slow of speech and of tongue.' Then the Lord said to him, 'Who has made man's mouth? Who makes him mute, or deaf, or seeing, or blind? Is it not I, the Lord? Now therefore go, and I will be with your mouth and teach you what you shall speak.'"

Isaiah 65:17–19 quotes the Lord speaking years before Christ's advent about his future creation—the same as the back bookend Revelation does,

God's creating new heavens and earth. "For behold, *I create new heavens and a new earth*, and the former things shall not be remembered or come into mind. But be glad and rejoice forever in *that which I create*; for behold, *I create Jerusalem to be a joy, and her people to be a gladness*. I will rejoice in Jerusalem and be glad in my people; no more shall be heard in it the sound of weeping and the cry of distress" (emphasis added).

• • •

The Psalms richly refer to God's creativity. Psalm 136 recites the refrain, "for his steadfast love endures forever." Verses 4–9 (omitting the refrain) highlight the creation: "to him who alone does great wonders . . . to him who by understanding made the heavens . . . to him who spread out the earth above the waters . . . to him who made the great lights . . . the sun to rule over the day . . . the moon and stars to rule over the night."

Psalm 146 exults, "Blessed is he whose . . . hope is in the Lord, his God, who made heaven and earth, the sea, and all that is in them . . ." (5–6). "[God] determines the number of the stars; he gives to all of them their names," says Psalm 147:4. Psalm 104:5: "He set the earth on its foundations, so that it should never be moved." This psalm expands upon many aspects of God's creation of the natural world, assenting, "O Lord, how manifold are your works! In wisdom have you made them all . . ." (v. 24).

Psalm 8:1, 3–4 acknowledges man's humility, given our small footprint in comparison with the rest of creation. This passage reads: "O Lord, our Lord, how majestic is your name in all the earth! You have set your glory above the heavens. . . . When I look at *your heavens, the work of your fingers, the moon and the stars, which you have set in place*, what is man that you are mindful of him, and the son of man that you care for him?" (emphasis added).

The Psalms echo God's creation of human beings in his own image, his highest creation, first learned in Genesis. "For you formed my inward parts; you knitted me together in my mother's womb," Psalm 139:13 says.

Psalm 148 urges God's praise from all of his creation, from heavenly hosts to human beings to beasts to all manner of earth and earthly creatures, on account of his creativity, seen in verse 5: "Let them praise the name of the Lord! For he commanded and they were created."

• • •

The New Testament continues to echo and affirm the Lord's role as Creator. There's no better place to look than the Gospel of John, chapter 1. This Apostle adopts the same phrase that opens Genesis: "In the beginning." Verses 1–3 of John 1 say, "In the beginning was the Word, and the Word was with God, and the Word was God. [Jesus] was in the beginning with God. All things were made through him, and without him was not any thing made that was made."

Various passages make brief reference to his creativity as a characteristic of God or as a fact. For instance, "mystery hidden for ages in *God, who created all things*" (Eph. 3:9), "for everything *created by God* is good" (1 Tim. 4:4), "such tribulation as has not been from the beginning of *the creation that God created* until now" (Mark 13:19), and "according to God's will entrust their souls to a *faithful Creator*" (1 Pet. 4:19; emphasis added).

In a discussion of divorce with some Pharisees, Jesus says in Mark 10:6, "But from the beginning of creation, 'God made [human beings] male and female.'" God's making human beings for intimate relationship becomes central evidence in Jesus's discussion of a human practice for the disintegration of marital relationships.

God's creativity appears in the book of Hebrews. In chapter 3, verses 3–4, we read, "For Jesus has been counted worthy of more glory than Moses—as much more glory as the builder of a house has more honor than the house itself. (For every house is built by someone, but *the builder of all things is God*.)" (emphasis added).

Hebrews 11:3 reads, "By faith we understand that *the universe was created by the word of God*, so that what is seen was not made out of things that are visible." Verse 10 speaks of Abraham's faith, making reference to God as the ultimate Creator: "For he was looking forward to the city that has foundations, whose *designer and builder is God*" (emphasis added).

Colossians 1:15–16 describes divine creativity and its purpose: "[Jesus] is the image of the invisible God, the firstborn of all creation. For *by him all things were created*, in heaven and on earth, visible and invisible, whether thrones or dominions or rulers or authorities—*all things were created through him and for him*" (emphasis added). This passage highlights the role in creation Jesus, the second Person of the Trinity, played. Jesus bears the image of the invisible Father in his human body. Jesus's divinity is seen in and affirmed by his taking part in the world's and the universe's creation.

Romans 1 reflects the fact that the Creator God puts his mark of authorship on his creations. Verses 19 and 20a read, "For what can be known about God is plain to [human beings], because God has shown it to them. For his invisible attributes, namely, his eternal power and divine nature, have been clearly perceived, *ever since the creation of the world*, in the things that have been made" (emphasis added). This disclosure of a creator's identity in his creations is reminiscent of Psalm 19:1: "The heavens declare the glory of God."

* * *

What has this brief exploration shown us about God as Creator? Scripture teaches God is eternal and infinite. He isn't limited, as we are. Though his creation is bound by time and space, he is not. He is able to speak creations into existence. His creatures reflect their Creator in various ways. His creations span a wide range of variety, scale, scope, and type. They include tiny things like atoms, molecules, and genes. They also include majestic mountain ranges, deep swirling oceans, and distant planets.

Further, Scripture provides examples where God avers his being the Creator of the earth, the universe, and their contents and inhabitants. Some passages contain human acknowledgment of God as Creator. Still other verses quote Jesus Christ, God's incarnate Son, stating God created all that exists. And, as seen in the passages from the Gospel of John and from Colossians, the preincarnate Jesus participated along with the Father and the Holy Spirit in the works of creation.

Finally, note something from the Genesis creation account. The unlimited God of the universe, who speaks things into existence ex nihilo, who is preparing a new, eternal heavens, earth, and Jerusalem, regards his creation as good. The first chapter of Genesis echoes a frequent refrain: "And God saw that it was good." The divine Creator, at the completion of his creation, looked at the whole of what he made and called the works "very good" (v. 31). This seems a fitting way to conclude one's creative efforts. This brings to mind Ecclesiastes 3:11: "[God] has made everything beautiful in its time."

Chapter 4
Made in His Image: Human Creativity

"The darling of courts in his childhood (for his father took him early on his travels for purposes of exhibition as a musical prodigy), the intensely industrious youth, the creator of a dramatic art in music, separate and by itself in the world, the greatest master of melody that the world has ever seen, the writer of innumerable symphonies, innumerable songs, innumerable sonatas, the possessor of a musical memory such as had never been conferred on the son of man before, [Wolfgang Amadeus Mozart] was the brilliant artist of high spirits, the man who lived life to the very last drop of the glass."
— Vernon Blackburn[1]

Imagine going with your grandfather into his workshop. It's filled with tools, hardware, various implements. The floor is half-swept; you spot vestiges of wood shavings, here and there washers and nuts, sundry scattered screws, pieces of bright copper wire, colorful clippings of electrical wire insulation. The florescent light over the divoted, scratched workbench illuminates a project half-finished, an open book, open owner's manual, a copy of *Popular Mechanics* revealing a dog-eared page, another project that's underway clutched in the vise.

Your grandfather shows you what he's completed: a small, rectangular box, evidently holding electronics, with clasps on the sides. There's a round grid of holes on one side. He says it's for your bicycle. Attach it to your handlebar and secure your cell phone (if imagining this scene takes you back mid- to late twentieth century, envision a transistor radio,

Walkman, or similar device). It will boost the volume of your music while making the sound crisper.

This labor of love for a grandchild is done by someone with the know-how and initiative to create a new electronic accessory for a loved one. There's the idea, the concept for putting the idea to practical use, sketches or other planning and perfecting aids. There's thinking through and tinkering with different shapes and sizes. There's trial and error with prospective materials, various ways of powering the device, figuring out how best to make the devices interoperate, and how to optimize each function. There also is consideration of costs, availability, and similar practical, nontechnical concerns. At its heart, this is an example of human creativity.

* * *

Here we examine a foundational precept: human creativity. Similar to the creativity that characterizes the Lord, he has endowed human beings with creativity. There's the critical distinction that God can create ex nihilo, speaking things into existence, and he can act by fiat in his creation, while human beings are finite and limited to discovering and working with things that already exist in nature or humans have created. But no other creature possesses this quality otherwise unique to the Creator God.

We start in Genesis, with the marching orders God gave the only creature with the divine spark. We'll see how the initial charge to his image-bearing creatures, the Creation Mandate, leads to human creativity and its workings. We'll consider key aspects of human creativity.

The Creation Mandate

Like God, human beings are creative. As God's image-bearers, we come by this characteristic honestly. He is creative; his highest creatures are creative. We are directed to labor, to use our brains, our brawn, and that creative spark among his wider creation. This order came in the Creation Mandate.

The Lord issued the Creation Mandate in the Garden of Eden. This command is broadly encompassing. It charges us with more than just reproduction of our species. This charge also calls us to use our creativity and to employ our dominion over the earth for productive and constructive purposes.

The Creation Mandate appears in Genesis 1, immediately after God created human beings:

> Then God said, "Let us make man in our image, after our likeness. And let them have dominion over the fish of the sea and over the birds of the heavens and over the livestock and over all the earth and over every creeping thing that creeps on the earth." So God created man in his own image, in the image of God he created him; male and female he created them. And God blessed them. And God said to them, *"Be fruitful and multiply* and *fill the earth and subdue it*, and *have dominion* over the fish of the sea and over the birds of the heavens and over every living thing that moves on the earth." And God said, "Behold, I have given you every plant yielding seed that is on the face of all the earth, and every tree with seed in its fruit. You shall have them for food." (vv. 26–29, emphasis added)

Our being idle was never part of God's plan. The Creation Mandate is all about people being fruitful, multiplying, and exercising dominion over the world. Being fruitful generally means forming families and flourishing by the usage and enjoyment of human faculties and characteristics, such as creativity, complementary design, and intelligence (or intellect)[2] and will.

Multiplying generally means bearing children within families, and thereby obtaining the inheritance God bestowed on humankind. By multiplication, married couples sustain and increase the human population. Procreation extends human presence around the world and continually replenishes and expands God's highest creature over the generations. God's design makes this possible. Human beings are created male and female, whose complementarity enables a man and woman together to have the capability of obtaining "the blessing of fruitfulness and increase."[3]

Dominion generally means responsibility to care for the lower orders of creation. God gave humankind dominion over every sort of animal, as well as over "every plant yielding seed" and "every tree with seed in its fruit" (Gen. 1:29). Before creating woman and before the Fall, "the LORD God took the man and put him in the garden of Eden *to work it and keep it*" (2:15, emphasis added).

• • •

We need to expand on *imago Dei* and its connection to dominion, as these form the realm where human creativity lies. Francis Schaeffer notes that human dominion over the rest of creation is based on our bearing God's image. "Dominion itself is an aspect of the image of God in the sense that man, being created in the image of God, stands between God and all which God chose to put under man."[4] The Divine Creator makes the image-bearer his agent over the rest of creation. We have dominion here. This means humans bear the privilege and burden of responsibility, serving as God's immediate caretakers over the rest of creation.

What does being God's image-bearer mean? That's a big question with far more depth and breadth than we can give it here. Matthew Henry names three facets of humans bearing God's image and likeness: "In his nature and constitution . . . of his soul"; "In his place and authority"; and originally "In his purity and rectitude."[5] Essentially, it means human beings are distinguished from all other classes of creatures by rationality, spirituality, emotion, love, and verbal communication.[6] "According to Scripture, the essence of man consists in this, he is the image of God. As such he is distinguished from all other creatures and stands supreme as the head and crown of the entire creation," Berkhof says.[7]

Humanity has the distinguishing ability of reason. Our rationality and verbal communication, for example, allow our comprehension from Genesis that human beings were made in God's image "in the flow of history," giving "an intellectual, emotional, and psychological basis to my understanding of who I am."[8]

This basis is rational, but also relational. Schaeffer observes that people are "created to relate to God in a way that none of the other created beings are."[9] Humans are related to God in a manner involving likeness and personal connection with the Creator. Like him, humans can communicate verbally, making personal interaction possible. Also, people possess emotion and love, just as do the three persons of the Trinity. And we share the quality of creativity, an aspect of our rationality.

We can recognize a crucial distinction between human beings and all other creatures, including the fruits of human labor. There are two basic categories among creatures: human beings and everything else. That is, people are God's image-bearers and capable of relating to God personally; all else Schaeffer refers to as machines. "Other things in the universe are properly [understood as operating] on a machine level: the hydrogen atom is a machine, the star system is a machine. Their relationship to God is mechanical."[10] As Psalm 19 tells, all the heavens and the earth glorify God in their mechanical way, while human beings do so in a special manner. It incorporates imago Dei, our relational capacity, rationality, spiritual dimension, and our creative, innovative capability.

We find the essence of bearing God's image, of humanity's divine qualities, in Ephesians 2:10. There, the Apostle Paul describes human beings as God's "workmanship." This word is followed by connecting what we are with God's creating us "for good works" in Christ he prepared for us to perform. We're his workmanship, and we're made to work and to "walk in" those efforts. Jonathan Master discusses our being God's workmanship, humanity's image-bearing of God's likeness: "The term used is *poiema* (the word from which we derive the English word *poem*). We are God's work of art, so to speak. . . . Rather, we are the workmanship of God, the Master Craftsman."[11] Human beings, then, are God's poetry, his work of art, unique within all of creation.

* * *

Consider the words used to describe God's purpose for placing Adam in the Garden: "to work it and keep it." This purpose survives human disobedience in eating the forbidden fruit, the consequences of the curse, and Adam and Eve's expulsion from the Garden.

"**Work**" relates to laboring. Other aspects of this word are nurturing, growing, making, and building. Work includes the idea of cultivation—in the Garden of Eden context, horticultural and agricultural cultivation. Phillips captures the meaning of "work" here as "bringing good things into being," "investing ourselves in accomplishing things of value," and, specifically for Adam, "making the original garden more fruitful and spreading its bounty in the world."[12] Our work mandate involves use of body and mind.

"***Keep***" relates to caring, tending, or protecting. *Webster's Unabridged Dictionary* lists fifty meanings of "keep." Psalm 121 provides examples of "keep:" "The LORD will keep you from all evil; he will keep your life. The LORD will keep your going out and your coming in from this time forth and forevermore" (vv. 7–8). For its part in the Creation Mandate, "keep" is related to "watch," "protect," and "guard."[13] We may also think of "keep" in the sense of "maintain" or "preserve" that which is in our care, such as a groundskeeper or shopkeeper. The keeper is tasked with keeping grounds in good order (e.g., lawn care, keeping the grass mown and the shrubs trimmed, raking smooth the sand in a bunker on a golf course) or keeping an eye out for shoplifting.

In line with working and keeping, Andy Crouch notes how God's planting the Garden of Eden, human beings' first home, models cultivation in "the combination of the beautiful and useful." There's ordering, working, and keeping, from his larger creation. In Genesis 2, God pays "attention to what already exists and what will be the most fruitful and beautiful use of it; most of all, what will most contribute to the flourishing of the human beings he is about to create."[14]

Notice how the Creation Mandate brings our imago Dei characteristics to bear upon our divine charge and responsibility to work and keep God's creation. The creatures made in God's own image reflect their maker intrinsically, as well as in the works they carry out and in the manner in which they work and keep. There is human creativity, bringing order to creation, and, in fulfilling their occupations, reflecting and bringing honor to the Divine Creator. And human labors produce fruit, literally or figuratively, a type of creation, of bringing about new things or improvements.

Thus, the creation of human beings coupled with the Creation Mandate integrate our imago Dei character with the work-and-keep directives given humanity at the beginning.

Human Creativity

I wonder what implements Adam invented back in the Garden of Eden. Surely, he must have identified a range of needs, problems to solve, better ways his tasks could be done. Maybe a comfortable contraption for sleeping. Tools for collecting the bountiful fruit in the Garden. Other things common to human life in both pre- and post-Fall states. After all, God charged Adam with working the Garden before his forbidden-fruit-eating disobedience radically changed things, among them making work more burdensome.

Wouldn't Adam have thought of ways to be more productive, to work more efficiently? Wouldn't some of Adam's new solutions or innovations lead to insights for making other improvements to or new applications of his earlier inventions? His working, keeping the Garden, filling the earth, and subduing plants, animals, and nature involved putting natural resources to practical use. After the Fall, Adam would have sought solutions and new implements to ease the more toilsome, more difficult working and keeping that fill human existence east of Eden.

• • •

How does human creativity work? Considering human creativity and our potential for innovation, what distinguishes people from machines may be summarized as our unique ability "to observe the world anew and make new discoveries about it." Pitted against four-year-old children in lab experiments, artificial-intelligence (AI) computing systems "can't seem to solve cause-and-effect problems. If you want to solve a new problem, . . . ultimately you have to experiment. . . . Innovation, even for 4-year-olds, depends on the surprising and unexpected—on discovering unlikely outcomes, not predictable ones."[15] For all its complex computational ability, AI seems bound to predictive analytics, not capable of truly original thought. AI is a sophisticated machine made by innovative human beings who, having God's likeness, possess the ability to discover the unpredictable, reason to solutions, and innovate.

In his book *Culture Making*, Andy Crouch discusses several qualities of human creativity that reflect the Lord's "creative character." One is **making something new**. Crouch notes that human beings always work "with the raw materials given us by God and by the generations before us. Culture is what we *make of the world*, not what we make out of pure imagination"[16] (emphasis in original). This is a pale reflection in humans (as the moon and stars are of sunlight) of God's ex nihilo sunshine, but a reflection of the Divine Maker nonetheless. For instance, "human language is so marvelously fruitful, linguists have asserted, that every human being who has acquired a rudimentary facility with language has uttered a completely original sentence: a combination of words that no one else has created."[17] Humans can, in a limited manner, bring forth something that didn't previously exist, from materials God made and man discovered, modified, or combined.

Another quality closely associated with creativity is ***imagination***. Ada, Countess of Lovelace, was well trained in and quite adept at mathematics. The daughter of Lord Byron calls imagination "the Combining Faculty. It brings together things, facts, ideas, conceptions in new, original, endless, ever-varying Combinations. It seizes points in common, between subjects having no very apparent connexion, & hence seldom or never brought into juxtaposition."[18] Lovelace, who refers to mathematics as "the instrument through which the weak mind of man can most effectually read his Creator's works," says imagination "is that which penetrates into the unseen worlds around us, the worlds of Science."[19] Imagination, the mental faculty for combining disparate things, would seem part of the secret sauce four-year-olds have for outsmarting AI in nonpredictive problem-solving.

• • •

Related to imagination, as Ada Lovelace defines it as "the Combining Faculty," is ***observation***. Creativity demands being observant—of nature, of people, of large things, of tiny things. Observation supplies data, stimuli, things human minds can process. We derive from that type of "data processing" patterns, insights, and ideas. The inventor of the telephone, Alexander Graham Bell, says, "God has strewn our paths with wonders, and we certainly should not go through life with our eyes shut."[20] Perhaps more to the point, beloved New York Yankees catcher Yogi Berra captures the role of observation in a Yogi-ism, "You can observe a lot just by watching."[21]

Human creativity has frequently involved observing nature. Artists and inventors alike have found inspiration, ideas, and insights from observing nature. Watching the brilliant colors of a summer sunset over the mountains, desert, or ocean; noting their interplay with clouds and the waning blue of the sky; taking in the last, angled, bright rays of sunlight. These may inspire an artist to commit a facsimile to canvas with oils.

Gazing and pondering the flight of birds sparked a wide variety of ideas in a wide variety of inventors for how human beings might soar aloft. Observation of natural phenomena has proven a rich, deep vein of sources of inspiration. Human creativity has drawn extensively to adapt what nature has to teach those humans willing to watch, learn, think, and imagine.

• • •

In addition, creativity involves **differentiation**. This is "paying attention to ordering and dividing what already exists into fruitful spaces . . ."[22] Crouch notes, "An essential part of the creative process is in fact the work of sorting, separating, and even excluding some alternatives in favor of others," which he calls cultivation.[23] Schaeffer refers to this as differentiation, such as God's dividing light from darkness, land bodies from bodies of water.[24]

Plantinga's description of differentiation (sorting, ordering, dividing, joining) provides rich insight into this important aspect of creativity on the divine side of the equation:

> According to Scripture, God's original design included patterns of distinction and union and distinction-within-union that would give creation strength and beauty. In Genesis 1 and 2, God—who sometimes speaks there in tantalizing first-person plurals that helped kindle early Christian interest in the distinction-within-unity of the Holy Trinity—sets about to dig out oceans, build hills, plant forests, and stock lakes and stream. But the setting for these endeavors is a "formless void." Everything in the universe is all jumbled together. So God begins to do some creative separating: he separates light from darkness, day from night, water from land, the sea creatures from the land cruisers. God orders things into place by sorting and separating them.
>
> At the same time God binds things together: he binds humans to the rest of creation as stewards and caretakers of it, to himself as bearers of his image, and to each other as perfect complements—a matched pair of male and female persons who fit together and whose fitting harmony itself images God.[25]

For us finite human creatures, our creative processes necessarily call for sorting, ordering, prioritizing, etc. along the way. "The best creativity involves discarding that which is less than best, making room for the cultural goods that are the very best we can do with the world that has been given us," Crouch says.[26]

• • •

Also, creativity is **relational**. The three persons of the Trinity create for each other the universe and all its creatures, Crouch notes. The Divine Engineer fashions each facet of his creation, building on what preceded it. Later creations relate to the foundational building blocks, such as light and dry land. God "created with a view to what comes before and after it."[27] In fact, there is much interdependence among various parts of the creation. "And in the climax of creation, it becomes plain that the whole world . . . is designed for the flourishing of excellently relational creatures, male and female, who themselves are very good because they bear the image of a relational God."[28]

In human creativity, the relational characteristic mirrors divine creativity in this regard "when it emerges from a lively, loving community of persons and . . . participates in unlocking the full potential of what has gone before and creating possibilities for what will come later."[29]

Many inventors are motivated to pursue a certain line of research or to make a certain invention by what it would do to improve things for a loved one or to meet a human need. Neurosurgeon and medical researcher Dr. Kevin Tracey was spurred to pursue medical research by his mother's death from a brain tumor when Tracey was five years old. He wished to spare other children the pain of that kind of loss. He veered toward the cause of septic shock that killed a patient, a baby girl who died in his arms. Dr. Tracey has invented medical technologies for stimulating the vagus nerve to control cytokine protein response in rheumatoid arthritis and Crohn's disease patients.[30]

French mathematician Blaise Pascal was motivated to invent a mechanical calculator, having seen the tedious labors of his father's job as a tax supervisor. He sought to alleviate the tedium. Pascal's mid-seventeenth-century invention "became the first calculator to be patented and sold commercially."[31]

Human invention often entails a community of specialists in a field. Many of them invent at the same time, together or independently, toward the same goal. Some may focus on one part of the larger goal, while others aim to solve all the relevant problems that must be overcome.

The Wright Brothers' interest in human flight grew in the 1890s, reading about bird flight and human efforts, such as those of Otto Lilienthal, who died in 1896 in a gliding accident in Germany. In 1899, Wilbur wrote and received from the Smithsonian Institution several papers on human flight and a list of books on the subject. As their pursuits in human flight increased, the Wrights met contemporary pioneers in the field, such as French engineer Octave Chanute. The independent-minded brothers—"who combine[d] mechanical ability with intelligence in about equal amounts"[32]—learned from the recommended books, including those by Chanute and by Smithsonian Secretary Samuel Pierpont Langley, articles and papers, and in correspondence with Chanute.[33] Such community, a relational aspect, provides inventors peers with whom to share, from whom to learn, and with whom to compete.

● ● ●

Thus, human creativity works by incorporating one's intellect in contemplation with one's abilities. This is tied integrally with one's relation to other people, and to the world God made and gave humankind to make something of. We use our intellect and our will, employing observation and imagination. We may express our thoughts in an art form, or we may apply them to a problem in need of a solution. Bottom line: Human creativity integrates the facets of our being in a concerted effort to make manifest original thought.

Chapter 5
Invention at Work

"Inspiration is never at variance with information; in fact, the more information one has, the greater will be the inspiration."
— George Washington Carver[1]

"A physicist is one who's concerned with the truth. And an engineer is one who is concerned with getting the job done . . ."
— J. Presper Eckert[2]

Remember "Necessity is the mother of invention."[3] Another way of putting it is human ingenuity steps forward when we face a challenge or see a need.

The prolific inventor Dean Kamen is an inventing wonder, akin to Edison's eclectic invention. Kamen invented the AutoSyringe drug infusion pump, the iBot all-terrain electric wheelchair, and the Segway, a two-wheeled, one-person, self-balancing electric vehicle, just to name a few inventions.[4]

Inventor Irwin Jacobs through his 1980s startup Qualcomm—now a Fortune 500 company—responded to the need for telephonic mobility. Jacobs and Qualcomm applied their knowledge of electronics and radio waves to address this challenge. They have successively extended the inventive process and application of science and mathematics to meet additional necessities.

As a result, today we have not only cell phones, but smartphones that put far more functions at our fingertips, such as GPS (global positioning satellite) maps, airplane mode, and app stores, than we could carry with us as separate devices. Of course, smartphones came about because of good old telephone service that Qualcomm's early cellular voice technologies helped make possible in the 1980s. Jacobs and company's semiconductors now set global technological standards and are found in all kinds of devices, implements, and equipment enabling or operating with wireless connectivity.[5]

These two men, like other inventors, see a problem (or rather, identify sequential challenges to solve) and apply their intellectual faculties and their scientific, mathematical, and artistic knowledge to working out viable solutions.

Jacobs and Kamen are inductees into the National Inventors Hall of Fame, among other awards and honors they've received. Their inventions—the fruits of their inventive labors—have met many needs. Their inventions have improved millions of people's lives and advanced human existence. Their inventions have benefited society in remarkable ways on a massive scale.

In short, these famous inventors have done as other inventors: They do what Plato summarized. A necessity drives them to create a solution to meet that need. This divine means of providing for God's creatures from the skills and abilities of other people, who derive benefits and blessings themselves from their own works, makes for an elegant formula.

● ● ●

We now turn to invention as it looks in practice. This examination builds on the general process of invention introduced in an earlier chapter, considering variation based on practicalities. As we'll see, invention and creativity rarely follow a clean, straight line. Illustrations from talented people who have pursued discovery and invention should make this discussion more concrete and clearer.

Necessity's Baby

How could their glider be controlled in midair? This proved one of the many challenges Wilbur and Orville Wright faced in inventing the 1903 aircraft that enabled human flight. The solution requires balance, "causing the center of pressure to coincide with the center of gravity."[6] This is hard to achieve because it depends on several factors, such as the shape and the angle of the wings and whether the craft is horizontal with the ground or at an angle. Design changes (e.g., increasing or decreasing the convexity of the wings) affect the center of gravity and the center of pressure, while changes in one cause shifts in the other.

The Wrights spent part of the summer of 1901 camped at Kill Devil Hills, their second year at North Carolina's Outer Banks, experimenting with a new glider model. This was no beach vacation. The Wrights' experiments left them bewildered and discouraged: lack of lift, lack of control with the front rudder, fierce resistance, and lack of speed downhill.[7] At the end of this trip, Wilbur and Orville "considered [their] experiments a failure."[8] The disappointments the Wright Brothers encountered led them to question the accuracy of published tables of pressures, on which data their glider designs had relied.

Back home in Dayton, Ohio, the Wrights designed and constructed a wind tunnel to test wing angle and other variables for balancing pressure. Their homemade testing contraption was made of wooden starch boxes (later replaced by a 6 foot x 16 inch x 16 inch box with a glass window), wallpaper scraps, and a bicycle wheel, with wind supplied by a gas-powered fan (later replaced by a belt-driven metal fan).

The brothers and bicycle shop mechanic Charlie Taylor also made instruments for measuring elements such as lift and drag. They crafted the measuring instruments from used hacksaw blades and bike spokes—"devilishly ingenious" and "phenomenally inexpensive."[9] The Wrights discovered German glider pioneer Otto Lilienthal's published table's angle was thirteen degrees too flat. Over several weeks that fall, Wilbur and Orville tested around fifty small wings of various shapes and designs. They followed these tests with "what was to be their crowning work of the year," several more examinations into angles of air displacement at ever more precise, quarter-degree measures.

In replicating the experiments of Samuel P. Langley, the head of the Smithsonian Institution, the brothers found their graphed data showed a hump at the 30-degree mark, in contrast with the smooth curve on Langley's graph. Upon examination, they discovered Langley recorded the same phenomenon at thirty degrees, but he had not plotted the anomaly on his graph. The Wrights thereafter would rely on the data as they exist, rather than Langley's misrepresented information.

Early on, the brothers read books, articles, and reports about animal and human flight. There was activity and inquiry around the world on the subject, some of it prominent and a good bit of it recent. Many enthusiasts and discoverers observed the flight of birds. Inventors in this field sought to apply the observations of winged flight. Some inventors used gliders, while others came up with various modes of powering aircraft of various designs. Some, such as intrepid glider Lilienthal, lost their lives in their pursuit of human flight.

The Wright Brothers, in 1899, watching pigeons in flight, noted how they changed the angle of their wings to turn and to restore balance. They noticed there was a relation between wing angle and air pressure, affecting control. "Thus the balance was controlled by utilizing dynamic reactions of the air instead of shifting weight," Wilbur put it.[10] This aspect of flight now had to be translated into practical embodiment for human flight.

A solution came to Wilbur one summer night in their bicycle shop. The older brother conversed with a customer to whom he'd just sold a bicycle tire inner tube. Wilbur pulled the inner tube from its rectangular box and, mid-conversation, "began absentmindedly to twist the ends of the open box in opposite directions. It suddenly occurred to him that if a frail pasteboard box could survive such strain, it might be possible to twist the cloth-covered wooden frame of a flying machine in the same fashion without sacrificing lateral stiffness."[11] Their solution to this aspect of accomplishing human flight the Wrights called wing-warping.

In these ways, the benchwork of the soon-to-be-famous Wrights informed new iterations of their aircraft, continual improvements of details and systems, additional tests of new models in their wind tunnel, advancing from glider to mechanized flyer. Two years hence, the Wright Brothers made history December 17, 1903, on the dunes of North Carolina. There Orville Wright piloted the first-ever manned flying machine that "raised itself by its own power into the air in full flight, had sailed forward without reduction of speed, and had finally landed at a point as high as that from which it started."[12]

Orville Wright flies the craft he and brother Wilbur (right) built, making the first successful self-sustaining flight ever. The Wrights' historic moment occurred on December 17, 1903, at Kitty Hawk, North Carolina. (Photograph by John T. Daniels. Library of Congress.)

• • •

As the Wright Brothers knew, invention involves **_applied research_**, applying knowledge to practical use. Invention is an iterative process—but it's not cookie cutter. This process integrates curiosity, insight, knowledge, adaptation, and dedication. An inventive effort is typically broken down into segments—for instance, designing and testing various designs or finding the most suitable materials.[13] It's an applied version of the scientific method, with the development and testing of hypotheses, replication and data generation, self-correction, and discovery of new knowledge on the way to creating a new or improved implement.

Remember Edison's quip about inventive genius being 1 percent inspiration and 99 percent perspiration?[14] Inventing involves a lot of trial and error. Insights may come only after expending great amounts of time and energy. Other times, insights pop into mind in an "aha" moment.

Educated guesses aren't uncommon. Usually, there's a combination of these modes of inspiration and progress, entailing inspiration along with mental and physical effort.

"Problem solving is pretty much why you invent," Qualcomm's veteran engineering executive Susan Armstrong says.[15] Her experience pursuing cutting-edge invention in wireless communication technology has required a lot of hard work and few "aha" moments, to achieve progress. "Certainly never happened to me," she says. Armstrong's big breakthrough came, not in "a light bulb moment," but in consistent application of her faculties, knowledge, and experience with ethernet and wired communications while serving on a project team working on digital fax for sending data. "My observation was that the voice link [of cellular telephone] is just a set of bits," Armstrong explains. "It's just a digital link. And so you can use that for data and for internet access."[16] Armstrong's "simple packet data" invention stemmed from her realization—"to me, the digital CDMA [code division multiple access] voice link looked like just another Ethernet, or dial-up modem link." This led to a demonstration of internet surfing on a Qualcomm mobile phone at a 1997 trade show.[17] Her invention was the gateway to 3G wireless data communications.

Others have examples of "light bulb moments." Xerox inventor Gary Starkweather proposed the laser printer and handled the work developing its optical and mechanical elements. A key idea came to him while he cut the grass.[18] Others succeed through trial and error. In 1969, chemical engineer Robert Gore was looking for a way to economize the supply of an expensive polymer used in pipe-thread tape. Slowly stretching it resulted in the material's snapping. Finally, Gore yanked it. The material "expanded into a porous but sturdy membrane" that led to the pliable material in Gore-Tex, the water-resistant material used in clothing and other products.[19]

Similarly, former Beatle Paul McCartney refers to all the music he heard over the years growing up as a budding musician and as a member of the Fab Four as "data." It "had gone into my very sophisticated computer, the human brain, had jumbled up . . . and somehow, as the dream, it just tumbled out with this song ['Yesterday']."[20] Among some of the music in his data bank McCartney names Nat King Cole's "Stardust."

Between Armstrong's and Starkweather's inventing experiences falls nineteenth-century inventor Thomas Blanchard's "aha" moment for his "Blanchard gun-stocking lathe." Blanchard had just made an improvement to a lathe with a "wholly original cam motion"[21] when he took on inventing a lathe to mechanize carving a gun stock. The stock of a rifle or shotgun "has a variety of subtle curves along multiple axes, with dozens of recesses and connection points." Thus, making a lathe to mechanically cut and shape a gun stock was a complex problem. One day, "the whole principle of turning irregular forms from a pattern burst upon his mind." Blanchard, walking on a road, began shouting repeatedly, "I've got it!"[22] Not only would this lathe, patented in 1819, mechanize gun-stock production, his invention solved a broader problem of how to carve irregular shapes mechanically.

Such examples from inventors and creators show what putting useful arts and fine arts knowledge to applied use looks like, how knowledge (and persistence) empowers invention and creativity.

• • •

To expand on how invention and applied research work, consider **basic research**, which builds the knowledge foundation that invention applies. A shorthand way of differentiating between pure research and applied research comes from the computer pioneer J. Presper Eckert: "A physicist is one who's concerned with the truth. And an engineer is one who's concerned with getting the job done."[23] Eckert was an electrical engineer. Basic research adds to the body of knowledge; applied research applies that knowledge to practical ends.

Those who pursue basic research seek to expand human knowledge about natural, mathematical, or other phenomena. They employ the scientific method and the instruments of science and mathematics, reason, and intellectual faculties to extend human knowledge. Basic research, or pure research, is often motivated by scientific curiosity. Scientists engaged in basic research seek to understand and explain natural phenomena. They formulate and test hypotheses, conducting experiments where hypotheses are proven or shown invalid. Experiments may attempt to replicate an experiment to validate a hypothesis by producing the same result.

Pure researchers develop theoretical explanations for certain relationships found in nature. They're tested through controlled, experimental conditions, with the scientific conditions, the tests, and the results publicly disclosed. Colleagues in the same field attempting to repeat the experiment provide insights from a type of specialist "crowdsourcing." Such fundamental research findings provide the theoretical framework to build upon. Theoretical understandings from basic research may be applied by the researcher or by others searching for a practical solution. Basic research may precede, coincide with, or complement applied research. That is, basic research may inform or accompany invention at various stages of the process of invention.

Take some examples of basic research. The nineteenth-century English laboratory researcher Michael Faraday worked at the Royal Institution for one of the leading scientists, Sir Humphrey Davy, beginning in 1813. Faraday, a former bookbinder inspired by Davy's lectures on chemistry and electricity, discovered chemicals such as benzene and gained insights into the liquefaction of gases. Benzene and other chemicals Faraday discovered proved key to the aniline dye industry. Nevertheless, Faraday, who eleven years later directed the Royal Institution's laboratory and was a fellow of the Royal Society, became better known for his discoveries in electromagnetism.

Faraday began, as a side project, lab efforts into conversion of magnetism to electricity, writing that goal in his lab notebook in 1822. After four attempts to change magnetism into electricity over the 1820s, Faraday "noticed a weird effect that provided the vital electrical clue" in August 1831.[24] He wrapped insulated wire around two sides of an iron ring. To one he attached a battery; to the other, a galvanometer, which measures electric currents. Faraday watched for electric current to show on the galvanometer side when the battery was attached to the wire on the other side. That didn't happen. Instead, Faraday observed short spurts of electrical current register on the meter when he attached or detached the battery. That is, moving the battery changed the magnetic force on the one side, resulting in electric current on the other.

Later experimentation affirmed that movement of the magnet changed its magnetic field, causing electrical current—"Hence, distinct conversion of Magnetism into Electricity," he recorded in his lab notebook.[25] Another experiment proved the continuous movement of a magnet, thereby continuously changing its magnetic field, continuously produced electrical current, as shown on the galvanometer. In the process, Faraday invented a dynamo, the first device for magnetic induction of continuous electrical current. These foundational discoveries from basic research gave the scientific world new understanding of the electricity-magnetism relationship and how one form of such energy could be changed into the other, along with a mechanism for such applied purpose.

* * *

Another scientist, 2018 co-winner of the Nobel Prize in Medicine James Allison, learned about T-cells, a kind of white blood cell important in the body's immune system, while in college. T-cells became a major part of his professional life, in which Allison has performed basic and applied research. His master's thesis analyzed the biochemical and serological characterization of bacterial asparaginases, which are enzymes that cause remission in leukemia cases.

Allison, whose mother died of leukemia when he was eleven years old, explains the self-correcting scientific method underlying basic research:

> A scientist is usually focused on interesting, and hopefully important[,] questions, and generating experiments to test hypotheses. As a scientist, it is equally valid to prove the hypothesis true or false. That's fortunate, because many of our hypotheses are wrong; in fact, if you are asking interesting questions, most of them are wrong. Being wrong can actually be a good thing, because the answers generated in disproving an incorrect hypothesis will help you and others to propose alternate hypotheses. Then, you go back to the lab to do more—hopefully better—experiments.[26]

* * *

Whereas theoretical scientists engage in basic research, inventors usually pursue applied research. Inventors apply mathematical and scientific principles, formulas, and laws—the products of basic research—to the creation of a tangible solution. That is, an invention is the practical embodiment employing underlying scientific and mathematical truths.

The application of knowledge typically occurs in a focused effort to come up with a tangible solution. Inventors develop a device, implement, or other product—often a labor-saving, productivity-boosting device or implement that may have commercial potential. The common denominator is each invention involves problem-solving, the application of basic and applied knowledge to a practical challenge, such as Faraday's dynamo.

Inventor and entrepreneur John McCorkle, who has eighty-six patents, early in his engineering career worked for the U.S. military on foliage penetration radar. He came up with a number of inventions on this project, such as a balun, for balanced-unbalanced electrical line-shifting used to drive radar antennas; the algorithm that forms the radar images of well-hidden tanks or missile launchers; and ultrawide and superwide band communications. McCorkle's first company, Extreme Spectrum, built the first chip set[27] that would accurately synchronize ultrawide band communications. "One of the inventions is not only that it can be done, but exactly how you put the transistors into an integrated circuit and accomplish this function," McCorkle says.[28]

His first patented invention made teletype work better with ham radio. "I reasoned how the ionosphere moved around and how it would change the frequencies that were coming in," McCorkle says. "And so I came up with a circuit that would automatically track . . . how [the ionosphere] was moving the two tones around so that, since there's always one tone broadcasting, either tone, whichever tone I had, I would use to track the ionosphere so that my circuit knew, exactly when this frequency shift came, where the other tone was going to be." This invention delivered previously garbled, now clear teletype messages over ham radio. This was a side project on McCorkle's personal time; nevertheless, the invention was technologically significant enough that the U.S. Navy licensed McCorkle's invention.

Problem-solving is the "fundamental driver" of invention, McCorkle believes. He explains, "Like on the teletype, when it comes in with a garbled message, you'd want . . . a nice, clean message. Even though you can barely hear the thing and make sense of it, you want your electronics to actually figure it all out and give you a clear message."

McCorkle works a problem in a manner "that stirs the creative juices." He elaborates on his approach: "The first step is deciding, is [this] solvable?

You have to have at least some sense that you're going to spend time working on something and have it come to fruition," McCorkle says.

> "The fertilizer is that time at the very earliest stages, where you're just throwing out all kinds of possibilities of what might work, how you might solve this. And some of those methods you may say, well, physics doesn't support that; it's never going to work. . . . And then you go to the next thing, and at some point you may come to say, this is just not going to happen. Or as you're stewing on it, you think, all right, there's no fundamental physics that's stopping it.
>
> "It's just a matter of you figuring out what the crux of the issue is. What's the heart of the matter that you can affect, what are the things you can't affect . . .? And then you limit yourself to the things that you can affect and you see how far you can push them.
>
> "So there's always a next stage right there; . . . you solve a problem to some degree, but that doesn't mean [it's] all the way to what you'd wish for. So part of it is you settle for being part way there, you just move the ball."[29]

● ● ●

Cephus Simmons was working as a radiologist assistant at the Medical University of South Carolina, performing an intussusception reduction procedure. Intussusception is when part of the intestine slips inside another part of itself. To reduce intussusception requires pressurizing the bowel to reverse the telescoping. That day, a problem kept recurring as the team worked on a five-year-old patient. What usually took twenty minutes took twice as long.

"[Any] time we get a leakage around the catheter, it deflates the colon, so all the air rushes out of the colon, all [the] fluid, and then you have to start all over again," Simmons says.[30] Complications may result. "The longer the patient has an intussusception, the more probable they are to have bowel damage to where they may have to have a bowel resection—a piece of the bowel taken out," Simmons explains.

The incident with the five-year-old frustrated Simmons, who knows the dynamics for danger. He looked for alternative catheter products to provide a better seal, but nothing on the market was satisfactory. So, Simmons began making sketches of designs for a more airtight seal.

Simmons developed a double-balloon catheter.[31] He says "it was just [the] one instance that got me involved in [invention]. I never thought I would be one of those that invented anything. My whole goal was to take care of my patients, and the opportunity opened up to where I needed a new product to take care of my patients. It was a product that wasn't available. So that's what led me . . . from medicine into invention . . ."[32]

The Cephus Catheter came on the market in 2019. Simmons, using his invention, soon performed the same procedure as on that five-year-old a few years earlier. This one was over in three minutes. The Cephus Catheter now has several more applications beyond childhood intussusception reduction. It's used for "colon cancer screening with a virtual colonoscopy," evaluating colon surgery sites for leakage, and with a barium enema for certain colon cancer patients. Plainly, Simmons has lived necessity being the mother of invention.

* * *

The invention of Gatorade stemmed from a conversation between University of Florida medical school professor Dr. James Robert Cade and former Gator football player Dewayne Douglas, who worked full-time at the university hospital as a security guard and part time as a U. of F. football coach. Over their daily coffee, the coach said fifteen of his players were hospitalized the night before from heat stroke and exhaustion. He added that his football players never urinated during a game. As was common, football players didn't drink fluids on the field. That got the kidney specialist's interest and sparked an idea.

The head coach okayed Dr. Cade's and his researchers working with volunteers from the varsity B team. The Cade team took samples of blood and sweat and temperature readings from the ten guinea pigs and ran tests. These data guided their making a drink to counter sports dehydration by "replac[ing] the electrolytes lost through sweat during intense exercise."[33]

The first hurdle was the concoction tasted awful. The players spit it out and refused to drink the elixir. Cade took his wife's suggestion and added lemon juice to the drink. "It was drinkable; it still tasted terrible," his daughter Phoebe Miles says.[34]

Field experiments proved the Cade drink highly effective. The first one was an intersquad game between the B team and the A team known as the "Toilet Bowl." Routinely, the starters beat the second stringers. Not in 1965. The B team, fueled with Gatorade, beat the varsity squad in the fourth quarter. The head coach ordered up Gatorade for the second field experiment: the matchup against Louisiana State University, which was favored to win. The Florida team, now drinking Gatorade, beat the LSU Tigers. Gatorade also helped lift the Gators over Georgia Tech in the 1967 Orange Bowl.[35]

Soon, Stokely Van Camp, a food company, licensed the rights to market Gatorade, and Dr. Cade's invention became a commercial product.[36]

● ● ●

Inventors, who come from all walks of life, commonly have tinkered as children, disassembling and reassembling machines, devices, or toys. Congressman Thomas Massie, an engineer with degrees from the Massachusetts Institute of Technology (MIT), did that starting in elementary school.[37] His childhood inventions included a self-watering flower pot, a G-force meter, a slot car lap meter, and a robot arm.

John McCorkle had access to capacitor testers, oscilloscopes, meters, and tools in his father's electronics shop. His father encouraged taking things apart and putting them back together. The kids did projects around the house like putting in air conditioning. John built speaker cabinets for a junior high science fair project.

Curiosity drives tinkering, exploration, discovery, and creating new things from materials at hand. Young tinkerers wonder what makes certain things work. Those who become inventors, either professionally or on the side, typically start coming up with ideas for solutions when they come face to face with a problem. Inventor Rob Yonover says, "I believe all people are inherently inventors and survivors."[38] Indeed, that aligns with God's giving human beings creative ability.

Our innate creativity activates as we contemplate how to solve a problem or find ourselves in the throes of a project or searching for an apt solution. As we've seen, creative people, such as songwriters and authors, and their STEM[39] counterparts may experience flashes of inspiration—an inventive insight, a melody or lyrics, a phrase, or an idea for something to draw or design. Many who've concentrated their intellectual faculties on something, even a term paper or a household project, have experienced the phenomenon of the "eureka moment." Astute creative types often keep pen and paper handy to jot down these flashes of brilliance.

Some solutions to problems before us come from nature. Many times, inventors adapt aspects of what they see in plants, animals, or other creatures. We saw this with the Wright Brothers' observation of birds' wings changing angles against the wind to effect changes in flight direction. Or one may gain insight and inspiration for an invention from reading about existing technologies. One may also conceive of an invention by mental acuteness and open-mindedness. Michael Faraday's colleague John Tyndall observed of his friend that when Faraday "attacked a subject, expecting results, he had the faculty of keeping his mind alert, so that results different from those which he expected should not escape him through preoccupation."[40]

Also, inventors keep a record of their ideas, their inventions, and their progress. Inventors' notebooks log important data, such as a detailed description of the invention and, often, diagrams. Inventors make notes on steps taken, materials used, results (both positive and negative), anyone else involved in the inventive process, source materials made use of, and the date of each set of information.

Yonover, who invented the RescueStreamer emergency signaling device, and Crowe suggest making a rough prototype early on to prove one's concept, before investing much in the effort. They consider prototyping in this manner a useful means of economically testing the viability of an idea for an invention.[41] It isn't unusual for inventors to go through several phases of testing, tweaking, iterating, and adjusting. An inventor may have to go back to the drawing board on some elements—finding the right material, the right design, or exacting other specifications, for instance. It's part of the process: Edison's 1 percent inspiration, 99 percent perspiration.

Once the various concepts of the invention come together, inventors build or commission a prototype suitable for presenting to investors and collaborators. At this point, an inventor has confidence that the concepts the invention embodies are viable; the underlying scientific and mathematical principles are rightly applied; and the selected materials, the engineering, and the design combine to show the invention will work.

As this discussion illustrates, invention is a labor-intensive form of problem-solving. Simmons explains, "Here's the overall problem and here's a potential solution idea. And then you have different . . . phases, material, or size or those kind[s] of things, . . . what the product is going to be used for, who it's going to be treating, and now to identify what material to use."[42] Simmons gives the example of product size: "Since I was doing it at first for intussusception reduction, I know that a small patient would take a 20-inch catheter. So I made sure I had a 20-inch [model]." He recalls having "8-year-old patients that needed [the procedure] and I could use a 30[-inch model] for those. So those were the two sizes I came up with because of the anatomical structure . . ."[43] Simmons notes that the thirty-inch model works well on adult patients.

Not only scientific factors weigh in the product development process, but also economic considerations. Inventor-entrepreneur Simmons tested market demand with a latex catheter, which costs an eighth of silicon's price to make. When he saw "that hospitals were demanding it, but wanted it in a nonlatex [material], we knew that we needed to put the money in to get the silicon version in [the] market so that we could supply it to all those customers that were requesting [the Cephus Catheter]."

• • •

Research and development, or R&D, companies—beehives of inventors—focus on advancing technological boundaries through cutting-edge invention. That means problem-solving on a grand scale. They rely on teams of scientists and engineers to leapfrog the current state of the art with new, improved practical applications.

Susie Armstrong relates an example from her work. "One of the big problems in cellular and wireless communications is efficient use of [radio wave] spectrum," Armstrong says. "And so that's the problem that basically gives rise to many of our most remarkable inventions: . . . How do you squeeze more throughput, more data, into . . . this very limited resource called spectrum?"[44]

Years after Armstrong and colleagues' efforts highlighted the need for faster, more spectrum-efficient technologies, Qualcomm researchers worked on making millimeter shortwaves of spectrum useful. Millimeter waves weren't practical to use for cellular signals because an object such as a tree or a wall breaks the signal. Another team was tasked with solving this problem for cellular communication applications. "Out of that came all kinds of inventions on . . . beam-forming and how you actually can steer around or compensate for . . . when [those] millimeter . . . waves hit something and disperse."[45]

That Qualcomm team accomplished utility of previously unusable millimeter waves for cellular wireless connectivity. Such projects often produce many inventions, some employed in the solution and others set aside for now. "[Within solving a problem], you get a lot of room for creativity, and necessity for creativity and invention [of] many, many different kinds. There's all kinds of inventions that go into how you make efficient utilization of spectrum," Armstrong says.[46]

This invention model may prove prolific such that some of today's discoveries and inventions aren't directly applicable to the project at hand but might be developed in the future. Armstrong notes, "You can look at something and say, I don't know quite what this is going to be used for, but it is interesting and it's likely useful . . ." Armstrong calls this "another way of using these new kinds of technologies. And I think that's a huge driver of invention, as well."[47]

* * *

Inventors have solved countless problems with new inventions and made countless improvements to existing technologies or devices. Invention in its many shapes serves as their problem-solving vehicle. They continue to invent, to discover, to improve the human condition, expand the economic pie, and advance human understanding of the secrets of God's universe.

Chapter 6
Inventors and Common Grace

*"[God] cause[s] the grass to grow for the livestock
and plants for man to cultivate, that he may bring forth food
from the earth and wine to gladden the heart of man,
oil to make his face shine and bread to strengthen man's heart."*
— Psalm 104:14–15

*"Enjoy life with the wife whom you love, all the days of your vain life
that [God] has given you under the sun, because that is your portion in
life and in your toil at which you toil under the sun. Whatever your hand
finds to do, do it with your might, for there is no work or thought
or knowledge or wisdom in Sheol, to which you are going."*
— Ecclesiastes 9:9–10

In September 2022, Hurricane Ian pounded Florida and South Carolina, with extensive weather effects in other U.S. states and Caribbean islands. Battening down and sheltering through a hurricane can be frightening, traumatic, and destructive. Human lives and private property hang in the balance of the fierce winds, rains, flooding, and other effects. It's enough to remind us of Noah.

God made a covenant with Noah following the global deluge that destroyed living creatures and mankind except for Noah, his immediate family, and the pairs of animals in the Ark. The Noahic covenant, found in Genesis 8:21–22, consists of God's promise never again to destroy all his earthly creatures, including humans, by water. He promises to maintain the environment, specifically seasons, the climate, and day-night cycles necessary to sustain life. This covenant provides the basis for the doctrine of Common Grace.

This covenant relates to God's general revelation, everything he provides his creatures in common. Common Grace involves God's pouring out blessings benefiting human beings, animals, and plants broadly in the here and now. The Noahic covenant doesn't involve humans' eternal redemption, as do other biblical covenants. Psalm 145:9 captures the gist of this doctrine: "The LORD is good to all, and his mercy is over all that he has made." Psalm 104, particularly verses 10–23, illustrates ways in which our good God bestows common blessings upon his creatures.

Common Grace extends beyond God's holding the whole world and universe in his hands. "Common [G]race blessings include sunshine, rain, food, and possessions (Matt. 5:45; Acts 14:17), wisdom or skill in crafts, trades, and learning (Dan. 1:4–5; I Kings 5:6; Prov. 30:1), family, and friends."[1] Common Grace is just that: grace in common. Matthew 5:45, for example, illustrates Common Grace: God causes the sun to shine on both evil and good people, Jesus-followers and those who don't follow Christ. He sends rain on the righteous and the unrighteous alike, watering their crops and satisfying their and animals' thirst.

Further, Common Grace extends to "a moral influence." To a degree, it restrains sin, keeps civil order, and feeds a level of righteous conduct within a society.[2] Romans 13 provides an example where Paul lauds civil government as appointed by God to restrain evil and to punish evildoers. He writes in the context of the imperial Rome of Caesars, not even of the former Roman republic epitomized by its Senate. Romans 2:14–16 describes the moral influence of general grace, saying people who "by nature do what the law requires . . . show that the work of the law is written on their hearts, while their conscience also bears witness . . ." Paul establishes in Romans 1:18–20 how humans can clearly discern God's moral attributes through nature, his creation.

Theologian Louis Berkhof describes how the Christian understanding of Common Grace came about:

> The origin of the doctrine of [C]ommon [G]race was occasioned by the fact that there is in the world, alongside of the course of the Christian life with all its blessings, a natural course of life,

which is not redemptive and yet exhibits many traces of the true, the good, and the beautiful. The question arose, How can we explain the comparatively orderly life in the world, seeing that the whole world lies under the curse of sin? How is it that the earth yields precious fruit in rich abundance and does not simply bring forth thorns and thistles? How can we account for it that sinful man still "retains some knowledge of God, of natural things, and of the difference between good and evil, and shows some regard for virtue and for good outward behaviour"? *What explanation can be given of the special gifts and talents with which the natural man is endowed, and of the development of science and art* by those who are entirely devoid of the new life that is in Christ Jesus? How can we explain the religious aspirations of men everywhere, even of those who did not come in touch with the Christian religion? How can the unregenerate still speak the truth, do good to others, and lead outwardly virtuous lives? These are some of the questions to which the doctrine of common grace seeks to supply the answer (emphasis added).[3]

Common Grace also means the laws of science and mathematics apply generally. The human aptitude to discover, understand, and apply complex scientific, mathematical, artistic, and technological truths isn't confined to individuals redeemed in Christ. Galileo Galilei, the Christian mathematician who discovered the Earth and planets revolve around the sun, wrote something translated as "Mathematics is the language with which God has written the universe."[4] Someone doesn't have to be a Christian to understand and use mathematics. (I can attest that being a Christian doesn't make you a math whiz.)

Applied science and technology remain grounded in the world that is. Scientists, mathematicians, and engineers must understand and employ the laws of science and mathematics to tangible or theoretical matters. Their work requires acknowledgment of limitations and incomplete knowledge in their field, and exercise of pragmatism as new information is found and new data generated.

This approach aligns with the biblical teaching "that God's Law is imprinted in the very structure of creation."[5] We live in "a world that answers to thought. Not, to be sure, one which we can hope to master with our finite minds; but that is our limitation, not the world's; for if it is a creation, and the product of perfect wisdom, it will be in principle intelligible."[6] Creators whose work is grounded in reality seek knowledge not as an end in itself, but for understanding their disciplines and nature, and as a tool for application in practical problem-solving.

In an earlier chapter, we examined God's endowment of human beings with qualities of creativity. These universally bestowed blessings make up part of Common Grace. The scriptural vine and fig tree metaphor,[7] like the Creation Mandate to be fruitful, applies to every human being in every age and everywhere, simply on account of every human bearing God's image.

Under Common Grace, every individual human being has dignity, whether he or she writes compelling literature, makes brilliant movies, invents prolifically, or suffers serious physical or mental infirmity. Therefore, we find both Christians and non-Christians possessing scientific, mathematical, and creative abilities by which they envision an invention or work of art and capably turn their ideas into tangible creations.

Some people are gifted with both scientific and artistic abilities. Samuel F. B. Morse, a Christian, was one of them. An itinerate painter, many of whose works were portraits, Morse's interest in electromagnetics resulted in his single-wire telegraph invention. Morse also developed the code that bears his name, used to transmit text with short and long electronic pulses representing letters of the alphabet. Morse had some success as an artist, and founded the National Academy of the Arts of Design in 1826. After praying about his election as the academy's first president, Morse accepted "the cause of the Artists . . . under Providence . . . in some degree confided to me."[8] Morse delivered a series of lectures that spring. Morse presented in one speech "a theology of the Fine Arts." This lecture discussed humans' image-bearing of the Maker and inability to create ex nihilo. "The artist can only combine into new forms the existent God-given 'principles' of nature, such as Motion, Order, Unity, and Mystery."[9]

Under Common Grace, society in general benefits from the fruits of the labors of such creative people. Every human being has a sin nature, but each of us bears the Creator's image. Some people have eternal life by faith in Jesus Christ. But that's separate from someone's creativity and abilities, which God has divinely endowed through Common Grace. Engineers, researchers, computer programmers, scientists, filmmakers, musicians, artists, writers, and sculptors each possess creative qualities, talent, and aptitude. These fall under the umbrella of Common Grace.

The Beliefs and Unbeliefs of Inventors

Human creativity and its fruits, under God's Common Grace, constitute one of his means of providing for and blessing his creatures. In this way, we're beneficiaries of inventors, whether they're Christians or not and whether we're Christians or not.

Moreover, inventors' creative gifts reflect the ultimate creativity of the Creator God, and even glorify him. This doesn't mean everything a human creates is good, true, or beautiful and glorifying of our Creator. Nor does it mean those with creative abilities intend their efforts to bring God glory. Some don't acknowledge God exists. Others merely believe in God. Some look to Jesus Christ for eternal salvation, even devote their lives and their temporal works to the Creator.

The electronics inventor John McCorkle credits the Lord and his faith in Christ as being crucial to his inventing. McCorkle can recount numerous instances in which he's "stewing on [a problem] . . . and I might pick up a magazine . . . [and see an article] . . . not talking a thing about the problem I'm working [on], but they're talking about a technique that they . . . used for doing something. And I'm looking at it saying, look how I can use this same concept and just apply it differently over here. . . . And it is such a God thing . . . it's like you're [asleep and] then awake." There's no doubt as to the source and provider of such timely, practical nuggets. McCorkle avers "these things that God touches you with that at the moment, at the time, it opens up a world that you weren't aware of before . . ."[10]

The point is Common Grace improves existence on earth for all humanity, regardless of an individual's standing with God, whether someone is an inventor or a beneficiary of an inventor's invention. Examples of inventors' and scientists' analytical abilities, ingenuity, and successes—some well-known, others less so—and their religious backgrounds will broaden our understanding of Common Grace.

Thomas A. Edison

Thomas Edison, the "Wizard of Menlo Park," held 1,093 U.S. patents. Among his many inventions were the incandescent electric lamp, the motion picture projector, and, of course, the light bulb. Edison increasingly focused on inventing in 1869, after working for Western Union as a telegraph operator. He regarded Michael Faraday a hero who shared Edison's modest background and found Horatio Alger–type success. Young Edison gained interest in electricity from working the telegraph. He read the three volumes of Faraday's *Experimental Researches in Electricity and Magnetism*.[11]

Edison's first big commercial break came when Western Union and financial tycoon Jay Gould bought the rights to Edison's invention, a quadruplex telegraph system, in 1874. Edison's technological improvement gave the telegraph company licensing his invention a faster system capable of sending four messages simultaneously over a single wire.[12]

In 1876, the young Turk Edison quit New York and Newark and set up R&D operations at a laboratory in Menlo Park, New Jersey, twenty-five miles outside New York City. His top associate, Charles Batchelor, and other mechanics, engineers, and skilled workmen conducted all facets of inventive discovery, development, and design. Indeed, Edison has been credited with "the invention of the method of invention."[13]

The Menlo Park facility, along with Edison's subsequent labs, became a model for industrial research and development. Together under Edison's leadership, the Menlo Park team brought forth all manner of inventions. Many of them, such as the phonograph for audio recording and playback and motion picture projection, proved foundational for new industries. Edison pioneered the electric automotive field. Other inventions of Edison's improved existing technology, as he did repeatedly with the telegraph and Bell's telephone.

Commercialization was top of mind for Edison. He assessed the commercial prospects of inventions to guide inventive pursuits. Patents were critical to his economic and commercial model. ". . . Thomas Edison was an exemplar of the patent licensing business model," law professor Adam Mossoff notes.[14] Edison licensed his patents to others, and the licensing fees funded ongoing R&D. "I always invented to obtain money to go on inventing," Edison explained.[15] Edison preferred R&D funded by revenues from his commercialized inventions. He also had to go to investors. "Up to the present time I have only increased the [battery] plant with profits made in other things, and this has a limit," he wrote Henry Ford. "Of course I could go to Wall Street and get more, but my experience over there is as sad as Chopin's 'Funeral March.'"[16]

Edison didn't shrink from the challenge of invention. He said, "I have not failed. I've just found ten thousand ways that won't work."[17] He invested long hours and energy into trial and error, observation and testing. At Menlo Park, he systematized and scaled the process of elimination and identification.

Edison also had an entrepreneurial bent. He got Western Union to help underwrite the Menlo Park facilities. And in 1878 he formed the Edison Electric Light Company, backed by big-time investors from the J. P. Morgan and Vanderbilt circles. Several years later, Edison Electric and its manufacturing arms became the conglomerate Edison General Electric, now called GE.[18] Edison General Electric centered on electricity generation and distribution systems where a central dynamo made power transmitted over wires to multiple locations. Edison adapted for electricity the means used to deliver gas for gaslights. Edison founded many businesses for commercializing his inventions. Not all of his enterprises succeeded, of course, and Edison didn't always make the most astute business decisions.

As for faith, Edison seems for the most part to have been something of a religious skeptic. One source calls Edison a man "imbued with devout Methodist beliefs."[19] However, in a 1910 interview with the *New York Times*, the great inventor dismissed that humans have souls, the existence of life after death, and eternal judgment. Edison referred to people as a "collection of cells" rather than individuals with body, soul, and spirit, insisting on a material, not a spiritual, existence. He referred to the human brain as "nothing more than a wonderful meat-mechanism."[20] Edison said he didn't believe in the "supernatural." He was skeptical of the spiritualism and occult practices popular at the time. However, his thoughts on the existence of electricity in both organic and inorganic things seem to have opened his allowing the possibility of some scientific explanation of psychic phenomena.[21]

There is a paucity of biblical spiritual fruit that might lead one to consider Edison a follower of Christ; however, there's no denying that many of his inventions were game-changers in the Common Grace realm. The fruits of Edison's inventive labors have enhanced the lives of billions of people for more than a century.

Michael Faraday

The humble scientist and inventor who rose from bookbinder to head of the Royal Institution laboratory, Michael Faraday made consequential discoveries and created a milestone invention, a dynamo. Ellen Vaughn writes that Faraday "believed that God had created the world and all its forces, and nothing ever disappeared from his creation."[22]

This confidence in the Creator and what Common Grace makes accessible to those who look closely lies at the center of Faraday's calling. His discoveries, such as that electricity could move a magnet and a magnet could induce electricity, led to electric-wire communications (telegraph, telephone); wireless communications (radio, cellular phone); and computers.[23] He observed promising electromagnetic properties in semiconductive materials.[24]

Faraday's faith in Christ informed his disinterest in fame and fortune, instead focusing his energy on discovery and invention. Faraday was "a devout member of the small Sandemanian Christian sect . . . [who] lived modestly, quietly, and happily with his beloved wife."[25]

His reliance on the Lord for his needs and to lead his course of research and discovery translates into Faraday's benefitting and blessing generations of human beings around the world from the nineteenth century down to the twenty-first century. Flowing from his Christianity and how God created him, ". . . Michael Faraday was a veritable lion, a passionate and brilliant scientist of rare energy able to select and focus on the most meaningful, discerning problems. His scientific output was prodigious and fundamental, influencing peers in many fields. His laboratory notebooks set a standard of beautifully observed detail, organization, and honest record keeping. The charm of his prolific writings—and his readiness to admit his many failures on the road to experimental success—earned him wide and enduring readership."[26]

George Washington Carver

George Washington Carver led the Agriculture Department of the Tuskegee Institute, to which Booker T. Washington recruited him. Together, they put Tuskegee on the map. Carver invented over three hundred products using peanuts (e.g., plastics, soap, cosmetics, dyes) and over one hundred uses of sweet potatoes (e.g., postage stamp glue, synthetic rubber, ink).[27] He became a well-known agricultural scientist, inventor, and widely respected expert. The Peanut Man obtained three patents, opting not to secure intellectual property rights to most of his inventions.

Carver, a Christian, attributed his inquiries, discoveries, and inventions to God. He said, "As I worked on projects which fulfilled a real human need, forces were working through me which amazed me. I would often go to sleep with an apparently insoluble problem. When I woke, the answer was there.

"Why, then, should we who believe in Christ be so surprised at what God can do with a willing man in a laboratory? Some things must be baffling to the critic who has never been born again."[28]

Born into slavery in Missouri during the War Between the States, Carver graduated high school in Kansas, earned bachelor's and master's degrees from Iowa State Agricultural College, and studied drawing and painting at Simpson College. He also homesteaded a plot of land on which he experimented with plants.

Carver's graduate research focused on plant pathology. This helped prepare him for development of crop rotation methods in the South, where King Cotton had depleted the soil's nutrients. The dirt became unproductive and, thus, of little economic value. Carver discovered how alternating production of other cash crops such as peanuts, soybeans, and sweet potatoes restored the soil's nutrients. He developed his many products from these plants in order to help spur commercial markets that would make growing the alternative plants profitable. Uptake of farming these nutritionally beneficial plants increased in the early twentieth century, particularly across the South; this improved economic opportunity for farmers, including poor, black sharecroppers.[29]

Closeting himself in his laboratory, Carver said, "Only alone can I draw close enough to God to discover his secrets."[30] Regarding divine inspiration in research and discovery, Carver said, "Inspiration is never at variance with information; in fact, the more information one has, the greater will be the inspiration."[31]

Carver relied on the Lord to direct his work, step by step, in his scientific and inventive pursuits. "God is going to reveal to us things he never revealed before if we put our hands in his. No books ever go into my laboratory. The thing I am to do and the way of doing it are revealed to me. I never have to grope for methods. The method is revealed to me the moment I am inspired to create something new. Without God to draw aside the curtain I would be helpless."[32]

Among his many honors and awards, the British Royal Society of Arts made Carver a member in 1916, and the National Inventors Hall of Fame inducted him in 1990.[33] Carver served on the Tuskegee faculty in Alabama from 1896 until his death in 1943.[34]

James P. Allison

Nobel laureate Jim Allison went from driven, talented Texas teen to premier medical research institutions such as the Scripps Clinic and Memorial Sloan-Kettering. In 2012, he began leading immunology at the University of Texas M. D. Anderson Cancer Center. *Wired* calls Allison "a cross between Jerry Garcia and Ben Franklin, and he's a bit of both, an iconoclastic scientist and musician known for good times and great achievements."[35] The harmonica-playing medical science researcher has sat in with country musician Willie Nelson, including on the TV program *Austin City Limits*.[36]

Allison's research efforts since the 1970s have led to a new class of cancer therapy. Allison identified how certain leukocyte antigens' proteins and T-cells prevent the immune system from going after cancer cells. Unlocking the details of this phenomenon, along with identifying T-cell receptors, opened the door to immunotherapy—inducing immune responses so the body can fight cancers through the immune system. T-cell receptors are key, as they distinguish among molecules.

Allison's research discovered the inhibitory molecule CTLA-4 was the culprit. If its receptor were blocked by an antibody, using what's called checkpoint blockades, the T-cell could do its healing work. He was able to bring the immune system to bear, prompting an immune response.

This led to cancerous tumors shrinking, even disappearing. Allison's breakthrough has led to development of a new class of biopharmaceutical therapies for cancers and autoimmune diseases. The first such medicine approved in 2011 for use in the United States, Ipilumimab, treats metastatic melanoma.[37]

Allison is decidedly a devoted Darwinist on origins. "I realized how fundamental the ideas of Darwin about selection was, [sic] fundamental to understanding biology, otherwise it is just a bunch of trees, flowers, species[,] all these stuff, family kingdom and all this."[38] While in high school, he took a summer biology class and worked in the lab at the University of Texas. Back home that fall, Allison declined to take a required biology class because it didn't discuss Darwin's theory of evolution. The compromise was Allison's taking a correspondence course from the University of Texas instead. Years later, he testified before the Texas legislature in favor of teaching evolution in public schools.[39]

Where is Jim Allison in terms of religious faith? "It comes down to a whole different realm, which is faith and belief. I am spiritual. I believe that there is something out there that was the origin of everything. I don't identify with any religion, but I respect that others do." Allison perceives God as similar to sunrise. "For me, God is something like that. It's just a feeling, something personal. The rest of the time I try to do more research on cancer to help people."[40]

Yet, Allison didn't hesitate when he had the opportunity to play harmonica with Willie Nelson, Mavis Staples, and Willie's band at a party, performing a gospel medley.[41]

Hedy Lamarr

Hollywood star Hedy Lamarr graced the silver screen in Metro Goldwyn Mayer (MGM) movies, casted with top-drawer leading men such as Clark Gable and Jimmy Stewart. The strikingly beautiful actress had a less-known inventive knack. Born Jewish in Austria, Lamarr took interest in how mechanical things work from her father's explaining various machines as the two took walks together. Thus, "at only 5 years of age, she could be found taking apart and reassembling her music box to understand how the machine operated."[42]

Lamarr converted to Catholicism to marry her first husband, who was ethnically half-Jewish. She left a bad marriage to a man who was close with Nazi and fascist associates, fleeing to France and then to pre–World War II London. There Lamarr met movie mogul Louis B. Mayer, who signed her to an MGM contract.

Appearing in such films as *Algiers* and *Samson and Delilah*, offscreen, Lamarr's life was more complex. She married six times, typically ending in bitter divorce. She raised her children Catholic. Her religious affiliation—and unspoken Jewish heritage—is likely tied to being Austrian while Nazi persecution of Jews was on the rise and later to portraying a certain image in her adopted United States, especially as a film star. Her children didn't know of their mother's conversion to Catholicism or Jewish heritage until after she died.[43]

"Improving things comes naturally to me," Lamarr said.[44] In her trailer, Lamarr kept a small inventing set-up, where she tinkered on ideas when she wasn't on the movie set, as well as maintaining an area for lab work at her home. She designed, for instance, airplane wings for boyfriend Howard Hughes, basing her innovations on bird wings and fish fins.[45]

Her most significant invention was a "secret communication system," a radio technology patented in 1942. She conceived spread-spectrum technology and, with the aid of music composer George Antheil, developed a prototype to demonstrate its efficacy. Lamarr and Antheil met at a party, where discussion of a German U-boat's torpedoing a ship evacuating British children to Canada led to further technological discussions and to collaboration. This invention enabled transmitter and receiver radio signals together to jump from one frequency to another. This "frequency-hopping" would lessen the likelihood of the signal being jammed by an enemy and allow a radio-guided torpedo to reach its intended target.[46]

Though the U.S. Navy opted not to implement Lamarr's innovation during World War II, by the Cuban Missile Crisis of 1962, a frequency-hopping system guided the torpedoes on U.S. ships. Years after Lamarr's patent expired, her invention has provided the technological foundation for modern wireless communications, reducing signal interference between users. This invention underlies cellular telephony, GPS, wi-fi, and Bluetooth wireless technologies.[47]

The tech-savvy Hollywood star worked on inventions all her life, several of them improvements on existing technology. For example, Lamarr invented a revolving chair for use getting in and out of a shower, a better traffic light, and a tablet that carbonated and sweetened water.[48] Lamarr continued working on inventive projects through the last years of her life, including a fluorescent dog collar and modifying the supersonic Concorde airliner.[49]

This impressive inventor, actress, and stormy-lived soul achieved fame on both sides of her talents. She's remembered and honored with a star on the Hollywood Walk of Fame and induction into the National Inventors Hall of Fame. Lamarr appears to have placed some importance on religious faith.

Hon. Thomas Massie

As he entered his teens, Congressman Thomas Massie says, he "was fascinated with my own arms and my own hands, and how they [differed] from most of the rest of the animals on the planet and how we had been endowed by God with the ability to change our environment, not just to be participants or victims of our environment. We are able to change our environment and make tools that let us change our environment even more."[50] The Kentucky Republican's fascination with human hands' capabilities set him on a path of discovery and invention of robotic arms and hands. That journey lasted through high school, undergraduate, and graduate study at the Massachusetts Institute of Technology: "There was this feedback loop where our hands allow us to invent, but the thing I chose to invent were hands. And then, by the time I got to college for my master's thesis, I decided to invent something that let you use your hands to touch virtual reality, to touch things inside the computer that didn't exist in real life. And so [I] went another level deeper in hands and touching and manipulation and interacting with the world."[51]

One of Massie's recent inventions is the Clucks Capacitor, an automated, mobile chicken coop. The commercially available invention moves poultry around a pasture by a solar-powered tractor, feeds them, and protects them from predators. The Clucks Capacitor may be controlled manually or by mobile app.

Massie believes God created humans in his own image and endowed "all of us with this desire to create." He contrasts human creativity with animal instinct. The honey bee's building a honeycomb and making honey isn't creative, but rather instinctual. Human beings "go well beyond what's required to survive when we create music and we create inventions in artwork. And I don't think you could explain that without appealing to a God having created this."[52]

The U.S. representative recognizes God's providential hand in his calling to elected office and becoming a leading advocate on patents. "When I ran for Congress, I had no idea that I would be dealing with patent law and patent policy, which is extremely shortsighted of me. I mean, I'm an inventor and I've dealt with patents my entire adult life. . . . I kind of have a unique role as far as patents are concerned."[53]

Raymond A. Wissolik

Raymond Wissolik was an electrical engineer who, at the television and electronics firm RCA, "designed and invented a number of vacuum tubes which found wide industrial use[,] including use in prototype color TV."[54] A Pennsylvania native, Wissolik's post–World War II career involved the U.S. Army Signal Corps veteran in exciting technological advances in electronic components, such as integrated circuits.

Wissolik, who earned engineering degrees from the University of Pittsburgh and the Stevens Institute of Technology (and a Harvard MBA), was named coinventor on a patent in 1957 for "Electrode arrangement of gas tubes." This invention for RCA improved thyratron tubes, which "were the early gate-controlled power switches, the forerunners of the Silicon Controlled Rectifiers (SCRs)," inventor Ron Katznelson explains in an e-mail. "The invention of the thyratron goes beyond its implementation in gas-filled tubes; it is about the principle of a 'sticky' switch—once turned on, it can only be turned off by removing or reversing power, which happens in AC power."[55]

Thyratron tubes were most recognizably found in dimmer switches for light fixtures as well as in electric heater controls and AC motor speed controllers. Katznelson says modern SCR light dimmers operate on the thyratron principle.

Wissolik was a member and elder of the Hendersonville, North Carolina, First Presbyterian Church. There Wissolik led Bible studies, taught Sunday school, sang in the choir, and ministered to people in need through his church's Stephen Ministry.

His great-niece Erica Wissolik is a Pittsburgh native who works for the professional association IEEE-USA and staffs its Intellectual Property Committee. Ms. Wissolik never knew her inventor great-uncle. Nor does she know about his religious beliefs or practices, other than family history and the obituary she discovered more than a decade after Mr. Wissolik's death. He appears to have been like many of the Greatest Generation, a devoted churchgoer and faithful Christian.

Raymond V. Damadian

In the 1970s, Ray Damadian invented the magnetic resonance imaging scanner (MRI), now widely in use for medical diagnosis. The New York City–born physician devised a means of mapping the insides of a human body with sufficient detail to differentiate between healthy and cancerous tissue.

He secured patent protection for the MRI in 1974, started a company to commercialize his invention in 1978, and successfully litigated against patent infringement. Damadian's successful invention won him a number of prestigious awards, including the Lemelson-MIT Lifetime Achievement Award and induction into the National Inventors Hall of Fame.[56]

Damadian, a Juilliard-trained violinist who finished undergraduate mathematics studies at age twenty, responded to the Rev. Billy Graham's Gospel appeal at Madison Square Garden the next year, 1957. Damadian's faith journey took him through wrestling with the secular tugs he found prevailing in academic science.

He eventually gained security in Christ and confidence in Scripture. Damadian came to hold to the Genesis creation account literally. As an inventor and scientist, Damadian says, "For me now the true thrill of science is the search to understand a small corner of God's grand design, and to lay the glory for such discoveries at the Grand Designer's feet."[57]

• • •

These and other inventors and scientists come from a range of faiths, even no faith. The variety of their religious views and their scientific, engineering, and other acumen illustrates Common Grace. Exceptional inventors and creators are found among Christians and non-Christians alike.

Common Grace means that being a talented inventor and one's religious faith aren't correlated. A non-Christian can discover and develop the secrets of T-cells or facets of radio waves and novel practical uses. A Christian may invent the MRI medical diagnostic device or discover the secret details of electromagnetism and a means of producing electricity.

Comparing inventors and scientists shows how fallen human beings reflect the God-given quality of creativity. Their stories show us how problems drive inventors to formulate solutions. These inventors' ingenuity illustrates how inventive solutions are an important part of creation and of God's Common Grace. And these accounts demonstrate how inventors resemble any subset of human beings—some are Christians, some are not. With Common Grace, people may benefit from each other. The creative fruits of their labors may improve the human condition for millions or billions of fellow human beings in this life.

Chapter 7
Creativity's By-Products: Human Flourishing

"Biblical flourishing encompasses all of our being, including our material, psychological, spiritual, and emotional aspects. The cultural vision of flourishing focuses primarily on our material prosperity with its false hope of happiness."
— Hugh Whelchel[1]

"Christian theology maintains the eschatological hope of a 'new heaven and a new earth' (Revelation 21:1): a promised future age in which the fallenness of the world is overcome, human existence is transformed, and God's good purposes for creation find their ultimate fulfillment. . . . But that is an eschatological hope: human flourishing in the present age will always be partial at best, subject to limitations, hindrances, and tragic conflicts between genuine goods."
— Neil G. Messer[2]

When we humans apply ourselves in creativity or ingenuity, we may get to a point where we feel drudgery or pleasure. Working out an original thought, a solution to a vexing problem, or an original device, we may well experience the latter—pleasure, fulfillment, a sense of accomplishment. Essentially, that's part of human flourishing.

The term *flourishing*, or thriving, covers a lot of ground. Some definitions will help us grasp its depth and breadth. "According to all traditional

Christian wisdom, human flourishing is the same thing as glorifying God and enjoying him forever, and human wisdom is an inevitable, and human happiness a frequent, by-product of such flourishing."[3] This example names flourishing's first principles and purposes: God is human flourishing's origin; applying our God-given abilities benefits us (e.g., wisdom, happiness), which glorifies God.

Andy Crouch defines flourishing "as fullness of being—the 'life, and that abundantly' Jesus spoke of. Flourishing refers to *what you find when all the latent potential and possibility within any created thing or person are fully expressed*. . . . I talk about the transition from nature to culture as a move from 'good' to 'very good.' Eggs are good, but omelets are very good. Wheat is good, but bread is very good. Grapes are good, but wine is very good" (emphasis added).[4] This definition engulfs the potential and the full expression, as the italicized phrase says, the raw material and the value-added creation. The examples of "good" to "very good" illustrate the fruits of human labor upon raw materials from nature. Here, flourishing is bringing forth something new derived from part of God's creation. It's tied to producing practical fruits from practical labors.

"Creation at its best leaves us joyful . . . ," Crouch says. "It prompts delight and wonder, even in the creators themselves, who marvel at the fruitfulness of their small efforts."[5] This part of Crouch's definition highlights something of what the creator or inventor derives from well-done work: joy, delight, wonder, marvel. These are the emotions children feel when they first see something like a lunar eclipse or first make something like a three-dimensional model. These feelings may not have been felt for a long time and, suddenly, one's artistic or inventive endeavor yields something better than we might have expected. How our hearts swell with delight and wonder at such times.

The beloved golf champion Arnold Palmer expresses flourishing, the sense of fulfillment "working and keeping" bring. "From a very early age watching my father I learned the value of a good day's work. It wasn't all about earning a living either. There is something satisfying about accomplishing something in a day, building toward something, creating something, or just putting your mind through some exercises of improving on a project or a task. I still get that satisfaction."[6]

Individual Flourishing, Vertically and Horizontally

Flourishing benefits an individual, while also affecting others. "In the Bible, shalom means *universal flourishing, wholeness, and delight*—a rich state of affairs in which natural needs are satisfied and natural gifts fruitfully employed, a state of affairs that inspires joyful wonder as its Creator and Savior opens doors and welcomes the creatures in whom he delights. Shalom, in other words, is the way things ought to be."[7] Flourishing, then, runs on both horizontal and vertical planes simultaneously. The human connects with the divine and with the human.

As individuals, we thrive from our creative effort, though even our greatest personal achievements and creative output don't satisfy our soul's deepest longings. They couldn't because our souls long for eternal completeness—to know the God who made us. Also, the process of invention is often iterative. As inventor John McCorkle says, ". . . there's always a next stage right there. . . . You solve a problem to some degree, but that doesn't mean it [is] all the way to what you'd wish for."[8] Our inventions and creative works are partial and temporal, creatures bound by time and space. Still, human creativity bears fruit, both tangible and intangible rewards.

Artist Claire Kendall flourishes as a Christian painter as faith "informs the way I look at subjects that I'm painting . . . :

> I see the beauty of God's created order. I see the beauty of the human face. I see the beauty of the human body. I see how complicated it all is and how amazing it is. And there is still a ton of mystery in it for me, but also I think [art done as unto the Lord] helps me to draw the beauty out of . . . things more, maybe because I see God, his workmanship. . . . Everything that I've painted is something that I find beautiful.[9]

In this manner, flourishing can be a sense of wonder in the midst of applying our abilities to our callings.

God wired us this way. The way things ought to be includes human beings flourishing, to God's glory and our own and our neighbors' good.

He provides us the ideal platform for being productive in a manner that can bring us fulfillment while blessing other people. His image-bearers have the tools we need to be creative—observation, imagination, intellect, verbal communication—courtesy of our Creator. We have the mandate of dominion, working, and keeping of the earth, plants, and animals. While Adam and Eve's Fall marred everything and made life harder, essential elements still function. Only now, those things are less efficient and more demanding.

Nevertheless, one person can accomplish quite a bit in this fallen world. Individual creativity and discovery, the creative works one person may produce, give us a beautiful picture of the tremendous potential lying within a single human being. It's amazing, but wait, there's more! Pull back the lens to a wider perspective, where we can consider the billions of people, each bearing God's image, and the broad, diverse sets of skills, interests, and aptitude God imparts to each of them.

The Cade Museum for Creativity and Invention has encapsulated this divinely endowed attribute of human creativity that produces flourishing as an intangible by-product of our ingenuity. Cade calls it "Inventivity®." "Inventivity is basically thinking like an inventor," Phoebe Miles says. "And . . . what does it mean? It's embracing problems without apparent solutions, recognizing failure is integral to success, and cultivating resilience, because taking a concept to reality requires resilience.[10] Miles adds curiosity about the unknown to the mix. When someone decides to adopt this attitude and undertakes an inventive effort, human flourishing follows, beginning with the creative individual.

Think of something you invented, created, or figured out that brought you a feeling of accomplishment. Now multiply that deep sense of fulfillment and whatever practical benefits may come from one person's creative endeavors by the number of engineers, mechanics, scientists, tinkerers, garage inventors, researchers, architects, artists, writers, film and television producers, directors, editors, and others, some with creative ability in the applied arts, others in the fine arts, some with curiosity, others facing an immediate need to find the solution to a problem. This close-up to wide-angle comparison gives a glimpse of the vastness of the potential for human flourishing. Thus, human flourishing explodes when the Creation

Mandate meets Common Grace. These divine blessings of God's highest creature help not only creators, but other people to thrive.

* * *

This broader sense of human flourishing isn't just potential, religious, or theoretical. Nobel laureate Edmund Phelps posits the "modern economy," epitomized and supported by extensive data from the Industrial Revolution and the following century, is characterized by economic dynamism. This dynamism occurs because of "indigenous innovation throughout the labor force."[11] In other words, the modern economy enables innovation—and human flourishing—on a massive scale. The average person's innovative capacity times multitudes of people tinkering and using their creative abilities yields innovations and tangible benefits for society. All that grassroots creativity is constantly being injected into a thriving, dynamic modern economy (as distinct from centrally planned or corporatist economies). "It is not an ordered system: It is turned topsy-turvy by homegrown innovation and the crazy scramble to create innovation."[12] Phelps expands on these subjects at length in his books *Mass Flourishing* and *Dynamism*.

The Columbia economist credits the West's, and particularly the United States', dramatic economic dynamism and its corresponding societal benefits to "the phenomenon of grassroots innovation by virtually all sorts of people working in all sorts of industries."[13] Fortunately for much of the West and the United States, their citizens have benefitted from certain underpinnings of society that facilitate human flourishing. These underpinnings include a free enterprise economic system, the rule of law, private property rights, and representative government. These things may strike some as normal. However, they are recent and exceptional. That such pillars of a nation are necessary supports what Phelps refers to as "endogenous innovation."

Phelps demonstrates that human flourishing by many, many human beings from all walks of life combines to multiply widespread invention and economic dynamism that benefits everyone. The unleashed potential of human ingenuity through free enterprise leads to untold innovative flowers blooming. Phelps preaches the need "to restore the grassroots dynamism that can drive transformational innovation for better lives—or what he calls 'mass flourishing'. . . that most innovation is not driven by a few isolated visionaries, but rather by dynamism on a mass scale: millions of people empowered to dream up, develop and market new products and processes."[14]

In this modern economic model, Phelps finds, "the idea that capitalism creates innovation rather than feeds upon it . . . explains why . . . countries with more inclusive institutions, irrespective of any other advantages or handicaps they might have, consistently have better records of long-term economic growth. It is because the inclusive countries *enlist a larger share of the national IQ, so to speak, into the scramble for innovation*" (emphasis added).[15] This is what the reality, beyond mere potential, of human creativity and discovery—human flourishing—looks like.

• • •

An example of human flourishing from Scripture illuminates an aspect of the Creation Mandate. It essentially calls humans to bloom where we're planted. Jeremiah 29:4–7 (NIV) echoes the Creation Mandate:

> This is what the LORD Almighty, the God of Israel, says to all those I carried into exile from Jerusalem to Babylon: "Build houses and settle down; plant gardens and eat what they produce. Marry and have sons and daughters; find wives for your sons and give your daughters in marriage, so that they too may have sons and daughters. Increase in number there; do not decrease. Also, *seek the peace and prosperity of the city to which I have carried you* into exile. Pray to the LORD for it, because *if it prospers, you too will prosper.*" (emphasis added)

In Jeremiah, there is definite instruction to be fruitful, multiply, work, keep, and take dominion, even in forced exile. We see that we are not only called, we're designed to use our human ingenuity to discover, to solve problems and make things and, in the process and as a result, to flourish, to thrive, to prosper.

* * *

As we've begun to see, beyond individuals' creative potential and realization, human creation extends flourishing more broadly. Creativity builds on what preceded it. Yesterday's discoveries, inventions, and other creative works contributed to the body of knowledge, technology, and culture of the present. Today's discoveries, inventions, and other creative works advance the body of knowledge, the state of the art of technologies, and culture. What's created today is improved and built on tomorrow. This steady enrichment flows from individuals to families to societies and cultures. It flows from generation to generation. This progression is another facet of human flourishing: a more encompassing type and scale of flourishing.

"Because culture is cumulative—because every cultural good builds on and incorporates elements of culture that have come before—cultural creativity never starts from scratch. Culture is *what we make of the world*—we start not with a blank slate but with all the richly encultured world that previous generations have handed to us," Crouch says.[16]

This link of human creativity to culture, along with its cumulative nature, means human creativity benefits society. Creativity-derived flourishing and its wider benefits occur because the Creation Mandate, introduced in an earlier chapter, commands humans to use their creativity. Harold Lindsell, in *Free Enterprise: A Judeo-Christian Defense*, calls the Creation Mandate "a charge to all human life to create wealth."[17] When we are creative and productive, working and keeping in God's creation, wealth and prosperity are economic fruits of human labor. Lindsell illustrates the meaning of the creation of wealth: "People produce more than they consume."[18] The abundance of fruits from human labor enables economic exchange, savings, investment, starting a business enterprise, helping the poor and infirm, and on and on.

Thus, flourishing stems from both the process of creating and the fruits one's labor produces. Multiplied by the billions of people who have lived over the millennia who have pursued new knowledge or its practical application, every iteration of making culture with every work of useful or expressive art—representing the combined creativity of innumerable individuals—becomes another brick in the massive mansion of human civilization.

These fruits of human creativity share human flourishing across time. We feel awe, inspiration, and the like when we examine 150- or 200-year-old invention models at the Smithsonian's Lemelson Center for the Study of Invention and Innovation, view masterpieces of painting by Michelangelo, Rembrandt, or Vermeer or hear the lovely, contrapuntal intricacies of the music of Bach, Corelli, or Handel. Museums, historical structures, libraries, historical reenactments—these treasure troves provide windows into elements of human flourishing that came before. We glimpse the needs, aspirations, and accomplishments of our human forebears.

These artifacts transcend time and space, just as well-written books, motion pictures, and images from the human imagination transport us in our minds to another place and time. A visit to a European castle can recreate the Middle Ages for modern visitors. A Christmas tour of George Washington's Mount Vernon makes the first president's home and the Washingtons' Christmas celebrations and traditions vibrant. The writings of Aristotle, Augustine, and Shakespeare convey the knowledge, wisdom, culture, and ideas of some of the best minds ever to live. How many countless souls have found comfort in the music and lyrics of such favorite hymns as "Amazing Grace" and "Rock of Ages"? Even just the first few notes of these tunes played on bagpipes, a piano, or a violin can touch listeners' heart strings, perhaps prompting tears to well.

Madeleine L'Engle, the author of *A Wrinkle in Time* and Newbery Award winner, explains a more profound aspect of human flourishing:

> In the beginning, God created the heaven and the earth. . . . The extraordinary, the marvelous thing about Genesis is not how unscientific it is, but how amazingly accurate it is. . . . Here is a truth that cuts across barriers of time and space.
>
> But almost all of the best children's books do this, not only an *Alice in Wonderland*, a *Wind in the Willows*, a *Princess and the Goblin*. Even the most straightforward tales say far more than they seem to mean on the surface. *Little Women, The Secret Garden, Huckleberry Finn*—how much more there is in them than we realize at a first reading. They partake of the universal

language, and this is why we turn to them again and again when we are children, and still again when we have grown up.[19]

We imbibers of creative works derive our own flourishing from a creator's or inventor's source of flourishing.

Nancy Pearcey describes this aspect of shared flourishing over time and space regarding Da Vinci's painting, *The Last Supper*. "The monk who sat at the [dining hall] table some five centuries ago, breaking bread and sipping wine, saw the solid wall give way and felt personally swept up into that dramatic moment of confrontation when the disciples asked, 'Is it I, Lord?' [who betrays the Son of God and puts Jesus on the cross]."[20] And today, viewers of this artwork may be affected as they gaze on Da Vinci's depiction of our Savior's last meal with his disciples before he carries out the mission for which he was sent as God in a human body.

These aspects of human flourishing, whether experienced by the creator, the consumer, or both, extend beyond merely the material. Rather, they encompass the psychological, spiritual, and emotional as well, which Hugh Whelchel says must be included in speaking of biblical flourishing.[21]

• • •

What else may we discern about human flourishing from this discussion? That the Lord endows human beings with creative attributes similar to his own. He gives humans dominion over his creation. We have a divine mandate to use our creativity, our intellect, and the resources he provides us, not only to survive, but to thrive.

Our creativity, coupled with our physical and intellectual capability, is to be spent for our own and our family's provision. We appropriately enjoy the fruits of our labor.

And yet, other people benefit from the blessings of what we create. Others benefit from the surplus we share. The person who invented the wheel is long dead, but all humanity continues to reap the benefits of that invention, while inventors continue to make improvements to that invention. In 1741, Handel poured himself into composing the oratorio *Messiah*, completing the work in less than four weeks' time.[22] The beloved music of *Messiah* has blessed perhaps billions of hearers over nearly three centuries and continues to do so.

Thus, human beings receive divinely given creative abilities and other faculties. Through their use, we flourish and thrive in various dimensions. We personally benefit while our creativity's spillover fruits benefit other people as well as God's creation. In various ways, our flourishing from human endeavor helps other people thrive.

Creativity's Benefits for Society

Every individual has some creative ability. Some people's ability lies in painting and drawing, others in musical performance or composition. Some can dance while others can act. Some are good at writing and others are good at mathematics or science. Some are more analytical while others synthesize information—each type leading to fresh insights and new knowledge. Some excel at applying mathematical and scientific knowledge to tangible or practical inventions.

What happens when individual creativity comes together? When creative individuals join their efforts? When people do what they're good at in cooperation with others who do what they're good at? A kind of human flourishing on steroids results. Andy Crouch notes that "while human flourishing is of paramount importance, the witness of Scripture seems to be that we human beings are here not just for our own flourishing, but for the flourishing of the whole created order."[23]

Of course, human beings typically take responsibility for their own needs and the needs of their families. Doing this is normal. Its essence is captured in a phrase related to work, industriousness, and economic endeavor: "to put food on the table." It's akin to "the fruits of your labor."

Sometimes, people may try to "boil the ocean." It's at best unrealistic and at worst utopian to set out to save "the world" and impersonalized humanity. It may result in neglecting one's family's needs and the needs in one's own community.[24]

The good news is, God has designed things so our individual and local efforts, including creative ones, may have derivative effects. Some actions may meet the needs of one's state, region, country, and others around the globe. Some creative output may benefit a subpopulation suffering from a particular disease or having some unique need. Another creation, such as a disease-resistant, genetically modified plant variety or a standard-setting microchip, may very well improve the lives of millions, even billions, of our fellow human beings.

Consider how individual creativity's fruits serve as building blocks for broader impact and greater good. We begin with biblically aligned principles God has made accessible through his revelation and Common Grace. Then we briefly look at how this works in practice, proceeding from individual to collaborations to institutions to societies.

The Solid Foundation and Scaling Up

Human nature, human characteristics, human limitations. We are finite creatures made by and in the image of an infinite Creator who highly regards creativity and ownership, even incorporating these facets into his design for his highest creation. Though divine image-bearing creatures, we also bear the consequences of Original Sin. These and other constraining factors have led to the discovery and development of concepts and legal principles such as property rights, the rule of law, free enterprise, and right of contract. God revealed these truths in the Ten Commandments and his law. When human laws, institutions, and practices related to creativity and ownership more closely align with divine principles, benefits follow for creative individuals, their collaborative partners, and people generally.

The foundational biblical principles of God's creativity and ownership of his creation helpfully inform a sound understanding of these qualities' endowment in human beings. This leads to incorporating his principles into human institutions and human creative and productive endeavors.

Secure ownership rights (which are discussed at length in Part II) enable people to save, invest, and be productive, inventive, and creative—that is, to be industrious and to apply their ingenuity. For example, with intellectual property protections of individuals' creative output and discoveries, individual creators can explore, discover, and make, secure in the knowledge that what they create will belong to them. Creating something that didn't exist before and owning what one makes incentivizes creativity and produces its by-product, human flourishing.

Economically exercising one's creative abilities and property rights in conjunction with other people exercising theirs is like joining bricks with mortar. Together, many individuals can comprise a corporation or an industry. Together, they can achieve much more profound breakthroughs and accomplishments than any individual could achieve alone.

Thinking about human flourishing in this context, individuals apply their talents in the research lab, on the factory floor, in an office, or a machine shop. Each one is part of something much larger than himself or herself. He or she contributes to producing things no one person could do alone.

Whether engineering the next-generation microchip that sets the standard for the Internet of Things or creating new and better antibiotics from genetic material, such endeavors involve substantial investments of time, money, and resources. They require a commitment to the long haul, through many dead ends and much trial and error. For each person involved in such an enterprise, this honest labor fulfills his or her God-given creativity. It provides not only financial reward, but a sense of satisfaction and pride in well-done work on something substantial and worthwhile. Such human endeavors draw from multiple individual and group incentives and yield a mix of individual and group benefits.

What does this look like? The next Thomas Edison may be talented at inventing things but isn't very good at turning inventions into products or getting products into customers' hands. Or an invention is sophisticated and will take a long time and much money to move from concept to commercial product. Mutually applying property rights enables making arrangements benefiting individuals and others.

With secure private property rights, the new Edison may coordinate with financiers, product development specialists, manufacturers, marketers, and others whose talents add value in getting a project executed and scaled up, the product made, distributed, and sold, and revenue generated. It enables him to build a business, raise capital, and invest in R&D.

The greater the economic value of each of his inventions or creations, the more important and valuable the property rights attached to them. They constitute the foundation of the superstructure. With this solid foundation, the collaborative enterprise-level or industrial-level floors of the structure will hold together. This is human flourishing on a much higher level.

Brick by Brick and Pretty Soon You Have a Wall

As an element of human flourishing, private property rights allow us to work with other people in collaboration within an enterprise. We can confidently combine talents and resources when we each have secure ownership rights. Each thriving individual in such a setting contributes to achieving things far exceeding the sum of their parts.

For example, the cancer drug-testing firm KIYATEC is an enterprise from which we can consider the kinds of human flourishing an invention and intellectual property ownership can yield. Doctors prescribe medicine for cancer patients undergoing chemotherapy. A patient may experience side effects from one prescribed medicine, while another drug would work without adverse effects. KIYATEC's technology accurately predicts which cancer medicine best suits a particular patient.[25] This kind of personalized medicine benefits patients because the drug that's the best fit for an individual maximizes the medicine's effectiveness while maximizing its safety. There's more gain, less pain. This specificity helps improve the clinical outcomes.

Adopted on a large scale, KIYATEC's predictive accuracy could save billions of dollars in medical expenditures. KIYATEC enables doctors to avoid having to treat adverse side effects on top of the cancer. Insurers would incur fewer costs due to personalized drug selection on the front end. The data enable pharmaceutical companies to understand their drug products better and share that information with medical practitioners and researchers. These examples individually constitute important benefits. Together, they represent significant aspects of human flourishing.

Now consider additional beneficial effects for individuals and for society. KIYATEC would earn more revenue. The company could scale up, create more jobs, and hire more people who develop more applications for its platform's use on other types of cancer. This leads to a wider variety of more patients suffering less and fighting cancer more effectively. Health spending is curbed without rationing care or imposing price controls.[26]

This kind of win-win advances medical science, saves the health system billions of dollars, expands the economic pie while providing paychecks that support KIYATEC employees' families and rewarding the firm's investors with solid returns. And all those such as KIYATEC's suppliers and distributors benefit from more business and revenue, which in turn benefit their business-to-business partners and employees. This new economic activity generates tax revenue for local, state, and national governments to provide services helping our communities, such as police, fire, emergency medical response, schools, and records of births, marriages, licenses, property deeds, and deaths. The private investors who earn nice returns gain greater means and confidence to invest in innovators, including and in addition to KIYATEC.

Insurance companies won't care if KIYATEC makes millions of dollars while saving them millions of dollars and delivering better patient outcomes. Only Karl Marx would begrudge executives', employees', and investors' earning a healthy monetary return in the process. These fruits come after years of research and development, entrepreneurial lean times, and surviving the "valley of death" by piecing together enough investment and revenues to keep the lights on through those lean times. Perhaps the firm goes public, launching with an initial public offering where a wide array of investors can help KIYATEC meet further success. Or maybe a large pharmaceutical or clinical laboratory company acquires the firm. Or maybe KIYATEC thrives as a privately held company. All these outcomes are good options—any of which beats going bankrupt, the fate of many small companies.

* * *

One little company, cofounded by a savvy bioengineer to commercialize the fruit of his inventive labor, has the potential to make a significant contribution to the advancement of its field of technology and benefitting society. There are thousands of startups and early-stage firms in all sectors of the economy working to bring about benefits—for themselves and for society—in their fields.

Also, powerhouse R&D firms such as Bell Laboratories, a Bell Telephone/AT&T subsidiary during most of the twentieth century, is a good example of large-scale discovery and invention that contributes to human flourishing. "Bell Laboratories' primary task was to develop the telecommunications equipment and systems manufactured by AT&T, but it routinely engaged in a vast range of other basic and applied research."[27] Established R&D firms, such as microchip innovators InterDigital,

Nvidia, and Qualcomm, biopharma innovators Genentech and Eli Lilly, and medical device firms Stryker and Boston Scientific, employ thousands of engineers and scientists who work in teams on research projects plowing new ground of discovery and invention. Their ingenuity focused on solving important problems in all the useful arts creates new property that drives expanding the technological and economic pies.

Such human flourishing relies on the powerful combination of creativity and ownership. This produces new knowledge, new insights, new practical applications, new inventions, new business deals and collaborations, and real-world labor- and time-saving benefits that help grow the economic pie and ease human suffering. Meanwhile, individuals thrive and bless their fellow human beings with the fruits of their labor. That's human flourishing's virtuous circle.

Part II:
OWNERSHIP

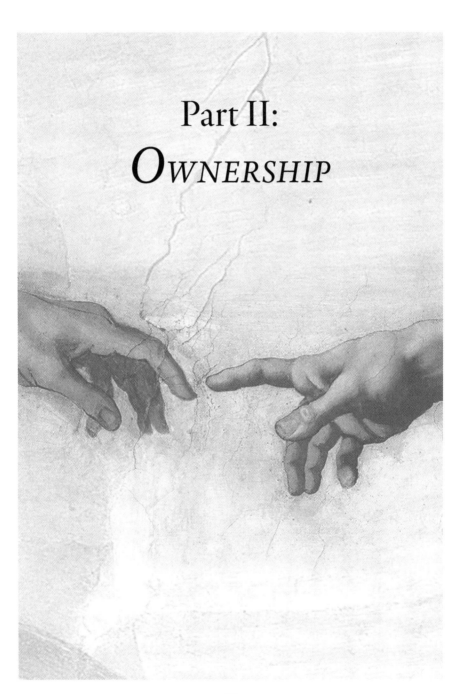

Chapter 8

God the Owner

*"The earth is [God's] by an indisputable title,
for he hath founded it . . . . It is his; for . . . He made it,
formed it, founded it, and fitted it for the use of man."*
— Matthew Henry[1]

Similar to creativity being inherent to the Lord, God inherently possesses ownership, or property rights. The Maker of all owns all that he makes.

We perceive in the Genesis creation account how God naturally owns everything he creates. This may seem obvious—so self-evident, in fact, you're thinking, *Why are you stressing this point? It's as clear as crystal, as plain as day. God made the heavens and the earth and everything in the universe, so of course he owns everything he created.*

I don't disagree; I simply ask: "If owning what one makes is obvious, then why do supposedly serious people purport to believe diametrically opposed philosophies such as communism and socialism?" The answer may be, "by their unrighteousness [they] suppress the truth" (Rom. 1:18). Further, it's imperative we proceed from the same framework, whether someone accepts it as true or rejects its veracity. This is true: God the Creator is also God the Owner of everything he has created. To move forward, this precept must be elucidated.

Biblical Affirmation and Evidence of God as Owner

As we saw with God's creativity, so too Scripture teaches the Lord owns what he has made. In Genesis, we read how God went about his creation of the universe. Here we pick up on his inherent ownership of his creative works. God never requests permission from anyone to make all he embarks upon creating. There's no consultation with any other being about any limitations, no getting prior input or permission to arrange creation as he wants, no one besides himself (the three persons of the Trinity alone) with whom to be concerned. He certainly never requires the input or permission from any of his creatures at any stage of the creation.

Psalm 24 declares God's ownership of his creation: "*The earth is the* Lord's, *and everything in it*; the world, and all who live in it; for he *founded it* upon the seas and *established it* upon the waters" (1–2 NIV; emphasis added). In fact, ownership stems from his having created these things.

Consider something else. "Then the Lord God formed the man of dust from the ground and breathed into his nostrils the breath of life, and the man became a living creature. And the Lord God planted a garden in Eden, in the east, and there he put the man whom he had formed" (Gen. 2:7–8).

What was God's purpose in creating human beings and in giving Adam charge of the Garden of Eden? "The Lord God took the man and put him in the garden of Eden to work it and keep it" (Gen. 2:15).

The Lord unilaterally decides to make a creature in his image, to plant a garden on a certain parcel of the Earth (2:10–14), to place the first human beings in that garden, to charge them with working the garden. Such command decisions imply ownership of each and every creature named.

Similarly, the Lord has every right, as owner, to impose conditions on his creatures. In Genesis 2:16–17, God designates the fruit of the tree of knowledge of good and evil off-limits. When this volitional creature encroaches this rule (Genesis 3), God the Owner casts his creatures Adam and Eve out of the garden: "Therefore the Lord God sent him out from the garden of Eden to work the ground from which he was taken" (v. 23). This sanction is perfectly within the Lord's prerogative as owner of every square inch of the universe and of every creature he fashioned and with which he populated his creations. God owns it all and may do with it as he wants.

• • •

The Ten Commandments provide additional evidence of God as owner. The commandments refer to ownership and assume private property rights—both for God and for humans. Notably, the Ten Commandments begin by prohibiting various forms of theft of God's property: "You shall have no other gods before me"; "You shall not make for yourself an image"; "You shall not misuse the name of the Lord your God"; "Observe the Sabbath day by keeping it holy" (Deut. 5:7–12 NIV).

Human beings are commanded to respect God's and other people's property. The first four commandments demand respect for the Lord's rightful place, befitting the creator and owner. This includes according him appropriate respect and honor through proper worship and Sabbath rest. They command our respect for his good name and forbid our appropriating it for our own misuse. These commandments resemble the principles of respecting various facets of people's property and keeping one's vows and contracts.

"The sum of the four commandments containing our duty to God, is, to love the Lord our God with all our heart, and with all our soul, and with all our strength, and with all our mind," question 102 of the Westminster Larger Catechism summarizes.[2] The commandments pertaining to God's property rights involve specific ways in which human beings are to respect his property, exemplify the order of love the Creator expects from his creatures bearing his image, and refrain from robbing God.

The prophet Malachi calls out the Israelites for stealing from God what belongs to him: "Will man rob God? Yet you are robbing me" (3:8a). In this context, the theft was withholding the tithe, but Malachi (and other Old Testament prophets) prophesies against many forms of robbing God. Reading Malachi and the first four commandments makes for an enlightening exercise, particularly when holding in mind God's being both the creator and the owner.

• • •

Elsewhere, Scripture affirms God's ownership of his property, his creation. For example, Psalm 103:19–22 says: "*The Lord has established his throne in the heavens, and his kingdom rules over all. Bless the Lord, O you his* angels, you mighty ones who do his word, obeying the voice of his word! Bless the Lord, all his hosts, his ministers, who do his will! Bless the Lord, all his works, in *all places of his dominion*. Bless the Lord, O my soul!" (emphasis added). This psalm resounds with elements of both the creation account and the first four of the Ten Commandments.

Psalm 95:3–7a reads, "For the Lord is a great God, and a great King above all gods. In his hand are the depths of the earth; *the heights of the mountains are his* also. *The sea is his*, for he made it, and his hands formed the dry land. Oh come, let us worship and bow down; let us kneel before the Lord, our Maker! For he is our God, and we are the people of his pasture, and the sheep of his hand" (emphasis added).

This passage leaves no doubt as to who created and owns the earth, its contents, and its inhabitants. That's one and the same Person. The Creator-Owner is referred to as a king over everything, including other objects of worship. A king by definition rules. He is sovereign. The God of the Bible is the God of gods, the superior over any and all comers. Here, his highest creation, human beings, is likened to sheep. We're ordered to "bow down," to "kneel."

In Deuteronomy 10:14, when Moses conveys to the Israelites what God told him to say, among other things he said, "Behold, *to the Lord your God belong* heaven and the heaven of heavens, the earth with all that is in it" (emphasis added). This comes as the Hebrew people receive the Ten Commandments and other divine statutes. Moses has exhorted the Israelites after their deliverance through the Red Sea and Pharaoh's drowning, and before Israel's journey to the Promised Land. Moses has already declared, "Hear, O Israel: The Lord our God, the Lord is one. You shall love the Lord your God with all your heart and with all your soul and with all your might" (Deut. 6:4–5).

This sober message encompasses much about God, his people, his law and commandments, and what he expects of the immediate Israelite audience. And in the middle of these instructions and a recapitulation of "what does the Lord your God require of you" (Deut. 10:12), Moses succinctly reminds the people the Lord owns all of creation. The juxtaposition of this sentence, about the God who has chosen them owning everything in the universe, strikes a contrast with the surrounding, extensive details of God's very specific commandments to this people. The command of their observance in love of God's laws is closely connected with God's creation and ownership of his people and everything throughout the universe.

At the end of the book of Job, the Lord replies to this righteous man who suffered severe affliction and testing. They aren't words of comfort, but words of tough love, answered "out of the whirlwind" (Job 38:1). "Where were you when I laid the foundation of the earth?" God asks (Job 38:4a). "Who determined its measurements—surely you know! Or who stretched the line upon it? On what were its bases sunk, or who laid its cornerstone, when the morning stars sang together and all the sons of God shouted for joy?" (vv. 5–7).

The rhetorical interrogatories continue through the following chapters. They cover many facets of creation, including darkness and light, rain and snow, constellations and various animals. Then at Job 41:11, God makes a direct statement of his owning all he has created: "Who has first given to me, that I should repay him? Whatever is under the whole heaven is mine." It's clear who is the superior, who is the Eternal Creator, the Divine Owner.

* * *

A recurring ownership analogy in Scripture is the potter and the clay. The creator has every right to determine what to create, how it's made, and its use. Isaiah 64:8 (NIV) says, "Yet, O LORD, you are our Father. We are the clay, you are the potter; we are all the work of your hand."

Isaiah 45:9–13 employs the same theme:

> "*Woe to him who strives with him who formed him*, a pot among earthen pots! *Does the clay say to him who forms it*, 'What are you making?' or 'Your work has no handles'? Woe to him who says to a father, 'What are you begetting?' or to a woman, 'With what are you in labor?'" Thus says the LORD, the Holy One of Israel, and *the One who formed him*: "Ask me of things to come; *will you command me concerning my children and the work of my hands? I made the earth and created man on it; it was my hands that stretched out the heavens*, and I commanded all their host. I have stirred him up in righteousness, and I will make all his ways

level; he shall build my city and set my exiles free, not for price or reward," says the Lord of hosts. (emphasis added)

This passage stresses the prerogative of creator and the inherent ownership of all a creator makes as well as the purposes for which a creator crafted those creations. Ownership flows directly from the act of creation.

In Romans 9, Paul recites Isaiah: "Shall what is formed say to the one who formed it, 'Why did you make me like this?' Does not the potter have the right to make out of the same lump of clay some pottery for special purposes and some for common use?" (Rom. 9:20–21 NIV).

We also learn of the maker being the owner from Jeremiah 18, which employs the potter-clay metaphor. Here God sends Jeremiah to observe the potter at work shaping clay. Then God says to Jeremiah: "'Can I not do with you, Israel, as this potter does?' declares the Lord. 'Like clay in the hand of the potter, so are you in my hand, O house of Israel'" (v. 6 NIV).

The connection between God's creating things and owning those things becomes firmer and clearer. From the Genesis account of *creatio ex nihilo*, to the Psalms proclaiming the earth and the fullness thereof as the Lord's property, to the potter having complete say-so—because he both makes out of and owns it—over what and how he fashions his creative works, to the Maker of the universe schooling Job with a description of creation that illustrates his majesty, sovereignty, glory, and power, an undeniable link exists between making something and owning it.

* * *

While Scripture speaks for itself—or more precisely, the Lord speaks for himself—it is helpful to us finite creatures to explore sound commentary on at least one key passage about God's ownership of his creation. Matthew Henry's commentary on Psalm 24:1–2 ("The earth is the Lord's, and everything in it; the world, and all who live in it; for he founded it upon the seas and established it upon the waters" [NIV]) covers this passage quite thoroughly and is worth our attention.

Henry notes Psalm 24's expansive scope included in its brief phrase:

> Here is, 1. *God's absolute propriety in this part of the creation* where our lot is cast, v. 1. We are not to think that the heavens, even

the heavens only, are the Lord's, and the numerous and bright inhabitants of the upper world, and that this earth, being so small and inconsiderable a part of the creation, and at such a distance from the royal palace above, is neglected, and that he claims no interest in it. No, *even the earth is his, and this lower world*; and, though he has prepared the throne of his glory in the heavens, yet *his kingdom rules over all*, and even the worms of this earth are not below his cognizance, nor from *under his dominion*.

(1.) When God gave the earth to the children of men *he still reserved to himself the property, and only let it out to them as tenants,* or usufructuaries: The earth is the Lord's and the fulness thereof. The mines that are lodged in the bowels of it, even the richest, the fruits it produces, all the beasts of the forest and the cattle upon a thousand hills, our lands and houses, and *all the improvements that are made of this earth by the skill and industry of man, are all his.* These indeed, in the kingdom of grace, are justly looked upon as emptiness; for they are vanity of vanities, nothing to a soul; but, in the kingdom of providence, they are fulness. The earth is full of God's riches, so is the great and wide sea also. *All the parts and regions of the earth are the Lord's*, all under his eye, all in his hand: so that, wherever a child of God goes, he may comfort himself with this, that he does not go off his Father's ground. That which falls to *our share of the earth and its productions is but lent to us; it is the Lord's*; what is our own against all the world is not so against his claims. That which is most remote from us, as that which passes through the paths of the sea, or is hidden in the bottom of it, *is the Lord's* and he knows where to find it.

(2.) *The habitable part of this earth (Prov. 8:31) is his in a special manner*—the world and those that dwell therein. We ourselves are not our own, our bodies, our souls, are not. *All souls are mine, says God; for he is the Former of our bodies and the Father of our spirits.* Our tongues are not our own; they are to be at his service. Even those of the children of men that know him not, nor own their relation to him, are his. Now this comes in here to show that, though God is graciously pleased to accept the devotions and services of his peculiar chosen people (v. 3-5), it is not because he needs them, or can be benefited by them, for *the earth is his and all in it*, Exod. 19:5; Ps. 50:12. It is *likewise to be applied to the dominion Christ has, as Mediator, over the utmost parts of the earth*, which are *given him for his possession*: the Father loveth the Son and hath given all things into his hand, power over all flesh. The apostle quotes this scripture twice together in his discourse about things offered to idols, 1 Cor. 10:26, 28. "If it be sold in the shambles, eat it, and ask no questions; *for the earth is the Lord's*; it is God's good creature, and you have a right to it. But, if one tell you it was offered to an idol, forbear, *for the earth is the Lord's*, and there is enough besides." This is a good reason why we should be content with our allotment in this world, and not envy others theirs; *the earth is the Lord's, and may he not do what he will with his own*, and give to some more of it, to others less, as it pleases him?

2. The ground of this propriety. *The earth is his by an indisputable title, for he hath founded it* upon the seas and established it upon the floods, v. 2. *It is his; for, (1.) He made it, formed it, founded it, and fitted it for the use of man. The matter is his, for he made it out of nothing*; the form is his, for he made it according to the eternal counsels and ideas of his own mind. He made it himself, he made it for himself; so that *he is sole, entire, and absolute owner, and none can let us a title to any part, but by, from, and under him*; see Ps. 89:11, 12. (2.) He made it so as no one else could. It is *the creature of omnipotence*, for it is founded upon the seas, upon the floods, a weak and unstable foundation (one would think) to build the earth upon, and yet, if almighty power please, it shall serve to bear the weight of this earth. The waters which at first covered the earth, and rendered it unfit to be a habitation for man, were ordered under it, that the dry land might appear, and so they are as a foundation to it; see Ps. 104:8, 9. (3.) He continues it, he has established it, fixed it, so that, though one generation passes and another comes, the earth abides, Eccl. 1:4. And his providence is a continued creation, Ps. 119:90. The founding of the earth upon the floods should remind us how slippery and uncertain all earthly things are; their foundation is

not only sand, but water; it is therefore our folly to build upon them. (emphasis added)[3]

Certainly, Henry plumbs the depths of these two verses, providing many insights and deepening our understanding of all that's entailed in the opening words of a familiar psalm. Moreover, both Henry's comments and what they imply intertwine divine invention and divine ownership. The two are inseparable. While a limitless Heavenly Father retains ultimate ownership over everything in creation, Henry explains how human beings share in the ownership. We'll explore that in the next chapter.

* * *

The principles of private property and ownership are important to the Lord, as seen in this brief survey of Scripture from both Old and New Testaments. God stakes his claim over everything he's created. The Bible makes clear God inherently possesses ownership rights over everything in his creation. This fact is boldly proclaimed, and God makes no apologies for it.

This point, especially ownership's connection with the Lord's creation of particular items, sets the basis for our discussion of ownership pertaining to human beings as we proceed deeper into Part II.

Chapter 9
Made in His Image: Human Ownership

*"Every man will sit under his own vine and under his own fig tree, and no one will make them afraid, for the L*ORD *Almighty has spoken."*
— Micah 4:4 (NIV)

"The right to private ownership of property means no less than that no one has the right to take it away from its owner."
— Harold Lindsell[1]

Have you ever heard a high-pitched child's voice shout, "Mine!"? Or maybe heard it in stereo as two kids tug on the same toy or bike? You don't have to teach children about ownership. It's inborn. They understand that the things that belong to them are theirs. We know what belongs to me is mine, and what's yours is yours. Even babies tighten their grip on their toy or spoon if you try to take it from them. Tug hard enough and they may squeal.

Something you own might have been a gift from someone. It could be something you saved up to buy. It could be something you made yourself, an original work from your own imagination or an appliance, an implement, or a device you assembled from a kit.

I can think of belongings in all these categories. For instance, my parents gave me a new six-string acoustic guitar for high school graduation. I still have that cherished gift to this day, and I can't imagine ever parting with it. I also remember, as a young child, saving my money for a Polaroid camera—my first "big-ticket" purchase bought with my own money. And I loved building models. I assembled and painted model cars, ships, planes, military vehicles, and an occasional human being (or classic movie monster such as the Boris Karloff Frankenstein) model. In junior high shop class, I built a gun rack (much cooler than the bird house option).

The point is, after receiving a gift, buying something, or making something, there's no question as to who owns it, except in certain extraordinary circumstances, such as receiving stolen goods. Similar to creativity being inherent in the Lord and in people, so too do humans inherently possess rights to own property, just as our Maker has ownership claims. The foundational precept of ownership is innate in human beings.

Ownership Rights in the Decalogue

A prime starting point for what Scripture teaches about humanity and ownership is the Ten Commandments. The commandments are all about private property rights. The Decalogue not only refers to ownership, as a whole and in each part, the Ten Commandments assume private property rights—for God and for people.

As seen in chapter 8, the first four commandments prohibit theft of God's property: "You shall have no other gods before me"; "You shall not make for yourself an image"; "You shall not misuse the name of the Lord your God"; "Observe the Sabbath day by keeping it holy" (Deut. 5:7–15 NIV).

The last six commandments require respect for other people's property in all its forms. Most directly, the eighth: "You shall not steal." But also: "Honor your father and your mother"; "You shall not murder"; "You shall not commit adultery"; "You shall not give false testimony against your neighbor"; and "You shall not covet your neighbor's wife. You shall not set your desire on your neighbor's house or land, or male or female servant, his ox or donkey, or anything that belongs to your neighbor" (Deut. 5:16–21 NIV). Each commandment addresses a different facet of privately owned property.

What does the Eighth Commandment prohibit? Westminster Larger Catechism question 142 answers:

. . . theft, robbery, man-stealing, and receiving any thing [sic] that is stolen; fraudulent dealing; false weights and measures; removing landmarks; injustice and unfaithfulness in contracts between man and man, or in matters of trust; oppression; extortion; usury; bribery; vexatious lawsuits; unjust inclosures [sic] and depredation; engrossing commodities to enhance the price, unlawful callings, and all other unjust or sinful ways of taking or withholding from our neighbor what belongs to him, or of enriching ourselves; covetousness; inordinate prizing and affecting worldly goods; distrustful and distracting cares and studies in getting, keeping, and using them; envying at the prosperity of others; as likewise idleness, prodigality, wasteful gaming; and all others ways whereby we do unduly prejudice our own outward estate, and *defrauding ourselves of the due use and comfort of that estate which God hath given us.* (emphasis added)

Suffice it to say, this commandment covers a lot of ground. As the italicized text states, the commandment even prohibits stealing from ourselves.

Notably, Jesus never overruled the Eighth Commandment against stealing. His teachings tacitly affirm private property rights. For example, the parable about the wicked tenant farmers in Luke 20 assumes the land owner who planted a vineyard and hired tenants to keep it for him is justified in seeking payment from his vineyard's fruit. The tenants steal from the landowner, squat on his land, and murder his heir. The tenants are clearly the bad guys in this parable because they abrogate rights of ownership. The tenants in the parable violate the Eighth Commandment, and they compound that sin by injuring the property owner's representatives, including fatal injury to the owner's son.

* * *

We see that God commands humans to respect God's and other people's property in its various forms—not only their physical belongings, but also someone's life, someone's good name, someone's vows or contracts. It's not a leap to infer we are also to respect someone's creations; the creative works and inventions of a human being belong to him or her.

Examples of Property Ownership Principles in Scripture

Our God-given rights of ownership may be referred to as private property rights. Ownership, or property rights, appear throughout Scripture. The Creation Mandate in Genesis is a general order where our Creator God puts lower creatures under human beings' charge. We humans take possession and share ownership with the divine, ultimate Owner. We make improvements to and come up with inventions and works from various things we own, helping us perform our work and keeping.

The Creation Mandate encompasses Adam's—and our—being charged with responsibility for lower creatures. Human beings are given the job of stewardship over God's creation. Our temporal rights of ownership follow from the mandate's vein. Psalm 115:16 says, "The earth [God] has given to the children of man." The responsibility placed on God's image-bearers means we make decisions and take actions that affect the lives and well-being of those things under our care. This divine bequest, delegation of responsibility to humans, amounts to ownership as a caretaking role during our time on earth.

Because the Creation Mandate was delivered before the Fall of mankind, ownership of the creatures under our personal responsibility is bifurcated: God is the ultimate owner of all of creation, including of us. And each human being holds temporal ownership of certain property.

* * *

In the biblical example of property rights found in the ownership analogy of the potter and the clay, the creator decides what to create and owns her creative work. Isaiah 64:8 says, "But now, O Lord, you are our Father; we are the clay, and you are our potter; we are all the work of your hand." Isaiah 45:9b asks, "Does the clay say to him who forms it, 'What are you making?' or 'Your work has no handles'?"

Scripture also emphasizes the maker being the owner of his or her creation. In chapter 18, Jeremiah gets an object lesson. Verses 2b through 4 say: "'Arise, and go down to the potter's house, and there I will let you hear my words.' So I went down to the potter's house, and there he was working at his wheel. And the vessel he was making of clay was spoiled in the potter's hand, and he reworked it into another vessel, as it seemed good to the potter to do."

God tells Jeremiah: "O house of Israel, can I not do with you as this potter has done? declares the Lord. Behold, like the clay in the potter's hand, so are you in my hand, O house of Israel" (v. 6). This analogy, making a broader point for the Israelites, likewise affirms the rights of ownership for human creatures.

The ownership of the clay (the raw material) and the creative work is never in question in these passages. If this was a commissioned pot, the end product still belongs to the maker until the transaction of sale is completed. The money or other compensation received in exchange belongs to the potter. If there were not clear ownership in this situation, the lesson would not hold up. As God owns his creations and a potter owns her creations, every human naturally owns his or her creative works.

※ ※ ※

Another Old Testament metaphor encompasses secure private property: the vine and the fig tree. First Kings 4 describes the peace and prosperity enjoyed under Solomon's reign, saying "he had peace on all sides around him. And Judah and Israel lived in safety, from Dan even to Beersheba, *every man under his vine and under his fig tree*, all the days of Solomon"[2] (24b–25, emphasis added).

Micah 4:4 says, "*Every man will sit under his own vine and under his own fig tree*, and no one will make them afraid, for the Lord Almighty has spoken" (NIV, emphasis added). Micah's usage comes in the context of the attainment of peace—shalom—of nations giving up war, turning their weapons into farm implements. And Zechariah 3:10 reads, "In that day, declares the Lord of hosts, every one of you will invite his neighbor to come under his vine and under his fig tree." Here the metaphor applies regarding prosperity and hospitality, the sharing of one's private property with others.

The vine and fig tree metaphor was a favorite of America's Founders.[3] George Washington invoked it often. For example, writing to Lafayette, Washington says, "At length, my dear Marquis, I am become a private citizen on the banks of the Potomac, and under the shadow of my own Vine and my own Fig-tree . . ."[4]

• • •

Other Scripture passages hit themes of property ownership. These include work and the resulting fruits: the creation of wealth. These together relate to human flourishing, which God purposes for his highest creation.

For example:

> Behold, what I have seen to be good and fitting is to eat and drink and *find enjoyment in all the toil with which one toils* under the sun the few days of his life that God has given him, for this is his lot. Everyone also to whom *God has given wealth and possessions and power to enjoy them*, and to accept his lot and *rejoice in his toil*—this is the gift of God. For he will not much remember the days of his life because God keeps him occupied with joy in his heart. (Eccles. 5:18–20, emphasis added)

This passage underscores how our work, the diligent application of human faculties and abilities, yields fruits from those labors, and those fruits have value and build new wealth. There is also fulfillment and enjoyment from such fruitful work. Moreover, the worker owns the associated wealth and possessions his toil brings. This far surpasses the philosophy that the person who dies owning the most stuff wins.

Psalm 103:15–16 places human potential and accomplishments in perspective. It says, "As for man, his days are like grass; he flourishes like a flower of the field; for the wind passes over it, and it is gone, and its place knows it no more." And yet, while the years of a life are but a moment in time, a person's life and work have value, both in divine image-bearing and in achieving and creating the things God puts it in our hearts to accomplish. Otherwise, why would the God who "has established his throne in the heavens, and [whose] kingdom rules over all" bother to forgive, heal, redeem, crown "with steadfast love and mercy," satisfy us with good, and renew our youth (vv. 19, 3–5)?

Before the Israelites entered the Promised Land, Moses delivered God's law and his messages about their covenant. In Deuteronomy 8:17–18, God has Moses remind them: "Beware lest you say in your heart, 'My power and the might of my hand have gotten me this wealth.' You shall remember the LORD your God, for it is he who gives you power to get wealth . . ." Matthew Henry comments on this passage, "The blessing of the Lord on the hand of the diligent, makes rich both for this world and the other."[5] We're reminded here of the divine source of our talents and abilities, as well as the gifts and blessings we receive. Human beings make something of the resources of this world and, from our labors create wealth which we own.

Scripture describes the just situation, a person's right to enjoy the fruits of his or her labors, which one rightfully owns. For example, Psalm 128:2a says, "You shall eat the fruit of the labor of your hands . . ." This is consistent with what we've seen to this point and will see below.

* * *

The New Testament also affirms private rights to own property. We've already noted one of Jesus's parables that relies upon private ownership rights, without any reproof by Christ. To be sure, Jesus, the Word made flesh, teaches not to insist on our rights in every circumstance. We're to live with a lighter grip on what we own in this life.

In the Sermon on the Mount, Jesus says relating to personal property, "If anyone would sue you and take your tunic, let him have your cloak as well" (Matt. 5:40). Similarly, Hebrews 10:34 reflects this lighter grip on temporal property and sometime withholding of our justified pursuit of legal property rights: ". . . you joyfully accepted the plundering of your property, since you knew that you yourselves had a better possession and an abiding one." Here, persecuted Christians let go of earthly property in favor of eternal, more valuable property holdings. This Christlike disposition does not contradict rights of private ownership. "The Bible is full of legal directives and principles for dealing with personal wrongs. This makes sense only if human beings have actual rights from the Creator."[6]

Another familiar lesson Jesus taught came in a parable about property and stewardship. Matthew 25:14–30 tells the parable of the talents. Jesus tees it up, rhetorically asking his disciples who will be considered ready for Jesus's return from heaven. "Who then is the faithful and wise servant, whom his master has set over his household . . . ?" (Matt. 24:45).

The parable of the talents is about a wealthy man leaving on a journey. He "called his servants and entrusted to them his property" (25:14).

He gave one servant five talents, the largest denomination of currency at the time. The owner gave another servant two talents and a third servant one talent. The two entrusted with the larger amounts of money each worked to increase the monetary holdings. They each doubled their master's money by their financial labors. The third servant did what was commonly regarded as a prudent means of keeping valuables safe.

Upon their master's return after an extended absence, each servant presented the results of his use of the master's property. The first two received the same praise: "Well done, good and faithful servant. You have been faithful over a little; I will set you over much. Enter into the joy of your master" (vv. 21, 23). The third servant's low-risk, no-reward handling of the single talent he'd been responsible for was met with, "You wicked and slothful servant!" (v. 26). He was berated for not even earning interest on the talent from the bankers. The master took away this servant's one talent, giving it to the one with ten talents. Finally, this servant was cast out of the household.

The parable of the talents is about faithful stewardship in readying oneself in life on earth to enter God's eternal kingdom. But applying the faculties, abilities such as creativity, and adding value to and wisely using the belongings God has entrusted to each of us in this life are all part of everyone's stewardship responsibilities.

• • •

The potter-and-clay metaphor recurs in the New Testament. In Romans 9:20–21, the Apostle Paul echoes Isaiah: "But who are you, O man, to answer back to God? Will what is molded say to its molder, 'Why have you made me like this?' Has the potter no right over the clay, to make out of the same lump one vessel for honorable use, and another for dishonorable use?" This total control of the potter over clay and what he or she makes out of the clay necessarily involves the potter's ownership. This whole scenario rests upon private property ownership, authority over the choice of what to make, the purpose for the vessel to be crafted, etc. And in this instance, it is a New Testament Christian invoking an example from an Old Testament prophet.

* * *

When Paul writes Corinthian Christians in 1 Corinthians 6, the apostle points out their hypocrisy relating to ownership rights. The Corinthians demand justice and their own legal and property rights when someone has defrauded and wronged them, while "you yourselves wrong and defraud—even your own brothers [in Christ]!" (v. 8).

Moreover, Paul has no issue with private property and legal rights. "Paul can criticize believers for wronging and defrauding each other in minor ways only by assuming that wronged parties have been sinned against, that they have certain rights granted by God that others have infringed."[7]

In Paul's same letter to the Corinthian believers, he addresses their failure to support Paul and Barnabas financially when they're present working to build that church body. The Corinthians have supported other apostles. This leads Paul to ask, "Who serves as a soldier at his own expense? Who plants a vineyard without eating any of its fruit? Or who tends a flock without getting some of the milk?" (1 Cor. 9:7). Paul and Barnabas have a claim on the fruits of their labor, even though they have willingly foregone it there.

In Acts 4 and 5, we see the generosity and concord of Christ's believers. This section recounts the early church, just after Pentecost, thousands coming to belief in Jesus as the Messiah, the Apostles Peter and John healing a lame man at the temple in Jerusalem, and their coerced appearance before the Pharisees and Sadducees. The rapidly growing startup body of followers of the crucified, dead, resurrected, ascended Savior at that point "were of one heart and soul" and "had everything in common" (4:32).

This wasn't a communist enclave. This wasn't a socialist cell or a hippie commune. Private ownership clearly remained. The lighter grip on temporal things, practiced voluntarily by individuals as they felt moved to relinquish something they owned, was a phenomenon seen in the early-stage Christian church as it scaled up in Jerusalem.

Take Barnabas, a Levite from Cyprus. He sold a field he owned, then gave the entire proceeds of the sale to the apostles (Acts 4:37). Other Christians with means acted similarly in order to meet immediate needs among the large group then present in Jerusalem (v. 34). These voluntary acts reflect the private ownership and hospitality meanings of the vine and fig tree metaphor.

Trouble came when Ananias and Sapphira sold property but, when presenting the money, apparently represented it as the full amount received from the sale. In fact, the couple had "kept back for [themselves] some of the proceeds" (5:2). The problem lay with the misrepresentation, with their claiming more credit than they deserved, not with their donating only a portion of the property sale. Peter makes this clear: "While it remained unsold, did it not remain your own? And after it was sold, was it not at your disposal?" (5:4). Thus, private ownership isn't frowned upon in the New Testament, by Jesus Christ, his disciples-turned-apostles, or the Apostle Paul.

• • •

Thus, both Old and New Testaments confirm God has endowed rights of ownership in human beings. From the Creation Mandate in Genesis to the Ten Commandments and the law to the Old Testament prophets. From Jesus's teachings to the apostles to the early church in Acts to the Apostle Paul and the epistles. Indeed, the ownership of private property, along with human ability to create wealth and God's prospering his people (individually as with Abraham and Job and corporately as with the Israelites and the Promised Land, described as flowing with milk and honey), is echoed and affirmed in the New Testament.

Like God's ownership, people's owning various property is a blessing, coming from the Lord. Unlike God's eternal, ultimate ownership, our ownership is limited. What we own, we hold temporarily. We're stewards of what the True Owner has entrusted to our care.

Property Rights and Biblical Principles

The vine and fig tree metaphor provides an excellent place to discuss private property rights derived from scriptural principles. Dreisbach notes the phrase's "implicit recognition of the right and blessing of peaceably possessing private property."[8] Indeed, Micah 4 uses the personal singular pronoun "his" before "vine" and before "fig tree." This points to individual ownership of certain property and the fruit this property produces.[9]

"[George] Washington used the phrase 'my own Vine and my own Fig tree' as if to emphasize the blessing of owning private property: the right to be left alone and undisturbed in the peaceable enjoyment of one's own property."[10]

As we have seen, Constitutional Convention Delegate John Dickinson (DE) employed the vine and fig tree metaphor precisely to mean private property rights: "A communication of her rights in general, and particularly of that great one, the foundation of all the rest—that their property, acquired with so much pain and hazard, should be disposed of by none but themselves—or, to use the beautiful and emphatic language of the sacred [S]criptures, 'that they should sit *every man* under his vine, and under his fig tree, and NONE SHOULD MAKE THEM AFRAID'"[11] (emphasis in original).

Dickinson regarded the biblical metaphor of the vine and fig tree as best capturing the meaning of "perfect liberty."[12] Dreisbach explains:

> The vine and fig tree represent contentment; that is, freedom from want and covetousness (covetousness is, of course, a terror to the tranquility and security of those whose possessions are coveted). This image is also a symbol of freedom from fear, especially the fear of war and rumors of war. More generally, the vine and fig tree motif represents the security to produce and enjoy the fruits of one's labor undisturbed by either lawlessness or the usurpations of the civil state.[13]

Remember, King George III's "long train of abuses" and "taxation without representation" provide the context for the Founders' understanding of threats to private property.

Thus, the Founders readily grasped the meaning of Scriptures prohibiting stealing in all its forms. They appreciated the contrast of the type of peace that enables individual human beings to work and make, to cultivate and harvest, to flourish in their labors, and benefit from the fruits of their labor. For this concept of liberty, the vine and fig tree provide an apt picture. It clearly entails solid private ownership rights.

The breadth of the vine and fig tree metaphor, coupled with the Eighth Commandment and other passages such as those we've explored, affirms private property ownership. Ownership is the prerogative of every human being, not just Christians. Let's take a brief look at how this "unalienable right" endowed by our Creator applies these biblical principles more generally to humankind.

* * *

Harold Lindsell, who was editor of *Christianity Today* magazine, penned a book titled *Free Enterprise: A Judeo-Christian Defense*. In it, Lindsell writes: "Those who think of property as something external to man, consisting of only material possessions, have a very narrow view of property . . ."[14] He names several forms of property. Life, free speech, peaceable assembly, and religious freedom each constitute a type of property. Due process and property in one's person in addition to her or his possessions properly count as private property. Of course, land, houses, automobiles, clothes, furniture, computers, jewelry, money, savings accounts, whatever you may have inherited or spent your hard-earned money to acquire, each of these kinds of possessions is also property.

Lindsell calls one's ideas his or her property. "Those of inventive mind have their ideas protected by patents. Do not new ways of doing things or new ideas which have never been thought of before belong to their inventors?"[15] Not only what someone creates, but the economic and commercial use of the invention counts as private property. "Surely if people have property rights in their ideas, then it follows that they must also be free to merchandise those ideas or even keep them to themselves as they please."[16]

The ownership God has bestowed on human beings, then, applies broadly. Just as the Westminster Larger Catechism answer we saw above illuminates a wide swath of what "you shall not steal" covers, unalienable rights to private property—what we own, possess, are due—cover wide species of property. I like Lindsell's description of private property ownership as "belong[ing] to the essence of life or reality. It is not something invented, devised, or conferred on men by other men. Like the law of gravity, it is a given in nature by the creative fiat of nature's God."[17]

* * *

So, the scriptural principles pertaining to private property and ownership are important to the Lord. He stakes his claim over what he's created.

He commands human beings to respect other people's property in all its forms. From this we learn we each inherently possess divinely bestowed ownership rights. And these inherent rights belong to every human being because we are God's image-bearers.

As we have seen, one aspect of ownership rights is enjoying the fruits of one's labor. This part of human flourishing involves benefiting from what your work and abilities produce. Flourishing and enjoyment of the fruits of our own labor are more than feelings of fulfillment or satisfaction. These benefits of work, keeping, and creativity intertwine with our owning the tangible results of our productive and creative efforts.

The "good" to "very good" examples (e.g., bread from wheat, omelets from eggs) not only demonstrate human creativity, but humanity's divine right of ownership. Isaiah 65:21 illustrates how human effort leads to results, which lead to benefits of ownership: "They shall build houses and inhabit them; they shall plant vineyards and eat their fruit." This passage all but uses the phrase, the fruits of one's labor.

* * *

As we've seen, God owns everything he has made, and God made everything. Psalm 24:1–2 declares God's ultimate and entire ownership: "The earth is the Lord's and the fullness thereof, the world and those who dwell therein, for he has founded it upon the seas and established it upon the rivers."

Yet, the Lord instills in humans the divine attribute of private ownership of property, as seen in Scripture. Such property comes in many forms—land, money, financial instruments, houses, cars, boats, businesses, personal possessions, our creative works, and more. These we own, temporally, as stewards.

Both creativity and ownership rights undergird our relationship to God and our relationship to other people. This is part of God's Common Grace to humankind. The Lord and each of us possess these qualities. God is the ultimate owner and humans his temporal stewards.

With the biblical foundation of both these God-given qualities—creativity and ownership—established, we now turn to the application of these qualities to creativity, invention, and intellectual property.

Chapter 10

The Mutual Reinforcement of Creativity and Ownership

"The patent system . . . added the fuel of interest to the fire of genius, in the discovery and production of new and useful things."
— Abraham Lincoln[1]

Say you plant a flower or vegetable garden in your yard. You battle the elements, bugs, deer, and squirrels. When flowers bloom or vegetables sprout, you gather them when they are ripe. The produce belongs to you (minus what the bugs and animals pilfered). Either way, part of the motivation for gardening or any other creative endeavor is the promise of what's produced on the assumption if you made it you own it. This horticultural example brings to mind Psalm 128:2a, which expresses the fact of the creativity-ownership combination: "You shall eat the fruit of the labor of your hands."

We have seen how human creativity reflects the creative characteristic of our Creator God. We grasp he has endowed human beings, his highest creature, with this attribute. Further, by God's Common Grace, humans have a natural awareness of individual ownership of property in its various forms, including one's own creativity and one's creative output. Together, these inborn human characteristics also are attributed divine attributes—creativity and ownership—that mutually reinforce one another. How they are mutually reinforcing, particularly in terms of incentivizing creative effort, is this chapter's focus.

Creativity and ownership mutually reinforce one another in many ways. For example, the combination incentivizes creative acts in seeking to find a solution to a problem. These two attributes combine to offer the promise of an individual's enjoyment of the fruits of his or her labor. A primary fruit of one's labor is the usage of one's solutions or creations. Another kind of fruit is a creation's economic benefits. Yet another incentive and fruit of human creativity and ownership is human flourishing. Inventors and creators derive a sense of satisfaction and fulfillment from acts of discovery and creation, and from seeing the fruits these produce.

The mutually reinforcing nature of these inherent qualities, which our Creator has endowed in his highest creation, involves a range of motives and incentives. This mutual reinforcement yields a variety of effects and produces a range of benefits, some practical, others less tangible or more personal. Also, this mutual reinforcement can lead to cumulative creative impacts that, for better or worse, shape individuals and shape culture and society. In places that protect private property rights under law, invention, discovery, and creative output advance scientific, practical, and artistic creativity and consequent improvement of the human condition.

But we are getting ahead of ourselves. First, we must consider the bringing together of the basic elements of creativity and ownership. We consider two analogies, each involving fire.

Flint and Steel, Fuel and Fire

In the Boy Scouts, you learn how to make a fire without matches. This mystery seems like a deep secret of the universe at first. But then you learn about the tools, a flint rock and a piece of steel, used to create sparks. Cause a spark near dry grass, pine needles, or other natural combustible matter, and you can build a fire for light and warmth at your campsite, as well as a means of cooking. There is no fire without both flint and steel.

This illustration of two components to perform a process causing a spark is useful in considering how creativity and personal ownership combine to spark creativity in an individual. The spark lighting a fire under one's creativity is caused by the bringing together of our two things, creativity and ownership.

Similar to the flint and steel causing a spark, creativity and personal ownership ignite an individual to creative endeavor. Sometimes this combination even sparks significant creative endeavor. Abraham Lincoln, the only U.S. president with a patented invention, captured this potent combination's dynamic effect. Lincoln said in his lecture on invention and discovery: "The patent system . . . secured to the inventor, for a limited time, the exclusive use of his invention; and thereby added *the fuel of interest to the fire of genius*, in the discovery and production of new and useful things"[2] (emphasis added).

Here, interest corresponds to private ownership of one's invention or discovery. Genius corresponds to one's creative ability. In Lincoln's description, ownership—and thus the right and freedom to enjoy the fruits of one's labor—stimulates human beings to apply their faculties to making a discovery, inventing something, creating something.

To change metaphors, this marriage of one's inherent creativity and ownership produces offspring: creative inspiration, intellectual and perhaps physical effort, insights and understanding, discoveries and creations. With fire, you need both flint and steel to get the spark that provides the ultimate goal, flames. With invention and creation, you need both creativity and the rights of ownership to get the intellectual spark that yields an invention or a creation. Both elements are required.

Only after you have both foundational elements do you reach a point where you are sufficiently motivated to work with certain materials and tools and apply certain scientific, artistic or mathematical principles to obtain the desired result. With fire, the fuel is a combustible material like dry straw and the spark is, of course, the dynamic input combusting the flammable material. Similarly, invention and discovery involve adding fuel (ownership rights) to the fire (creativity) in Lincoln's metaphor to fan the flames and heighten the fire's intensity. In either instance, the combination leads to a roaring fire, with all its benefits.

Understanding the interdependence of our creativity and ownership, we consider aspects of how these qualities are mutually reinforcing. We do so through examination of the incentives to create.

Mutual Reinforcement: Incentives to Create

"Necessity is the mother of invention."[3] This saying adapted from Plato's *Republic* indicates one of the incentives to discover and invent. Indeed, finding a solution to a problem may well be the prime incentive for invention and discovery in most cases. However, several types of incentives could lead someone to pursue a creative endeavor. These include practical incentives, economic incentives, and expressive or personal incentives. Though more than one incentive may lie behind a particular creative effort, we will consider the types of incentives separately.

Practical Incentives

A practical need or problem at hand that requires a repair or solution drives people to be creative in search of a solution. This has been common throughout human history. In early historical times, somebody created the wheel. This invention is thought to have originated around 3500 BC in Mesopotamia. The earliest wheels were wooden with a hole in the middle for an axle.[4] Human beings have since found countless uses for wheels, from transportation to implements such as the wheelbarrow to the spinning wheel.

Throughout history, human beings have applied their creativity in countless ways to solve problems. The common factor is necessity incentivizing invention. Farmers, blacksmiths, craftsmen, and others have been motivated to come up with a new device or tool for performing a task or making some necessary step or process easier. There's a problem right in front of you. You work to solve that problem. You need a solution, so you devote your efforts toward finding a solution. When you come up with a solution, you put it to use in dealing with your problem.

An example comes from Thomas Edison's inventing the light bulb. The Wizard of Menlo Park envisioned a full electrical system to compete with coal-gas lighting. His laboratory worked to solve complex problems of the several elements of his direct-current (DC) electrical system. These included dynamos to generate electricity, wires to convey electric current, and electric light bulbs to illuminate homes, streets, and buildings. Ignoring the skeptics by pursuing incandescence, Edison endeavored to find the right combination of materials and specifications to make a filament that would glow, not incinerate. Thus, one challenge was to find a material for the filament, the part inside a light bulb that lights up when electric current flows through it.

Edison determined that, rather than low-resistance material, incandescence would require a high-resistance material. In early 1879, Edison was using platinum, but it wasn't long-lasting. Short bulb life wasn't practical enough. By October, his researcher Charles Batchelor turned to baked carbon and noted "some very interesting experiments on straight carbons made from cotton thread."[5] Such material in horseshoe shape proved to illuminate for fourteen and a half hours, a vast improvement on the already amazing two- to three-hour life of earlier versions of bulbs with the same filament material. After further experiments on other candidate materials for the filament, Edison and his team settled on carbonized cardboard. With this discovery and a better vacuum inside the bulb, they could move on to overcoming other challenges in their quest to have a working electrical system ready to display on the fast-approaching New Year's Eve.[6]

Bringing creativity to bear for solving problems isn't confined to Edisons but takes place throughout society. It often benefits others. In medicine, doctors and nurses use simple and sophisticated tools to diagnose, test, treat, and operate on patients. These include diagnostic equipment such as magnetic resonance imaging (MRI), electrocardiograms, and stethoscopes—each of which someone invented. Meanwhile, frontline medical practitioners may identify a need or get an idea for a medical technology they wish they could use with their own patients. They have an idea of something that could help their patients and doesn't exist presently. Some, like Cephus Simmons, will invent it.

Many times, the practical incentive in developing an idea for a new device or tool is to meet an immediate need that must be met before further progress is possible. For instance, a mechanic may have to develop a stronger material, reinforce a weak part of an implement, or fashion a better design to repair a broken tool or device. Other times, the practical incentive is saving time, easing a task (preserving energy or stamina), or achieving greater efficiency. Such practicality has led to the invention or improvement of untold numbers and kinds of labor-saving devices.

New technology may be both labor-saving and productivity-boosting. The common denominator with practicality-driven invention is each case involves problem-solving. It's practical and need-driven. Yet, without the natural consequence of owning the solution one develops and gaining its usage, there would be little incentive to invest one's effort into coming up with a solution. Without ownership, why waste the time and energy?

Economic Incentives

Deriving economic value from one's creative output is a common, indeed a very important, incentive for pursuing creative endeavors. The economic incentive is Lincoln's "fuel of interest." Whatever you invent may benefit you financially. The promise of a financial return on your investment of your time, talents, and resources makes the application of ingenuity worthwhile in a tangible way. But it's more than simply a payday.

For many, economic incentives may be the driving force for invention. An economic motivation for invention and creativity constitutes one way of enjoying the fruits of one's labor. Here, aside from the practical use of what you have invented, exchanging the creative output for compensation is a part of commercializing the invention. Now you've added a financial transaction for the new creation.

Thomas Edison was economically motivated. "I always invented to obtain money to go on inventing," the famous inventor said.[7] It was said of Edison, "He asks himself when a new idea is suggested, 'Will this be valuable from the industrial point of view? Will it do something important better than existing methods?'"[8]

Monetization as a fruit of one's inventive or creative labors represents one measure of the value of the creator's time and efforts. Only, rather than exchanging time and effort for wages, this economic exchange involves a valuable good derived from time and effort traded for money, a precious metal, another form of currency, or something else of value. That is, the newly created property holds intrinsic value. It springs from someone's intellect and other resources invested into creating the new intellectual property (IP), having no guarantee either the idea can be translated into something tangible or the invention can be successfully brought to market.

Assuming successful invention and commercialization, the exchange represents a fuller measure of the value of the product of intellectual effort.

It differentiates the thinking and conception of an idea from the physical aspect of an idea turned into something useful and tangible. The added value of the result of the entire process—for instance, a marketable invention—associated with the item reflects the difference between a one-for-one exchange and return on investment. The latter is similar to earning interest on deposited money in a savings account that compounds and grows.

Say you invent something of particular importance or value while working for your employer. You might receive a bonus or royalty share on top of your salary. Salary or wages compensate your time and labor. Further, the financial return to the employer on the employee's ingenious breakthrough on the firm's behalf may exceed that of mere commodities or consumer items. That's because the benefit to the buyer of a high-value, IP-based good is typically of much greater value.

As this important differential in value and return from IP sinks in, those with aptitude in mechanical, scientific, and other useful arts may aspire to intellectual labors rather than to simple physical labor. The potential reward from knowledge-based work helps make the risks of the former worthwhile. This aspect of economic incentive to create lays the groundwork for expanding the return on such investments and spreading prosperity. As with Edison's central question above, the prospective returns on economically driven discovery and invention fund research and development (R&D), capital investments in buildings and equipment, and creating jobs and hiring people for the pursuit of innovation.

For someone initially motivated by practical need, the creative solution may seem the end of the matter. You've solved your problem. You have your fix. But it may occur to you that other people face the same problem you had or have the same need. At that point, the incentive begins to shift from practical to economic.

This discussion brings to mind Jesus's parable about the talents (Matt. 25:14–30). The sophisticated manner of economic exchange we've discussed and the fruits of one's labor were understood at the time of Christ. How else could a parable be credible or even relatable in which the good servants could achieve a return of 100 percent on a specie of currency? A tenfold rate of return would be acceptable to many twenty-first-century venture capitalists who invest in R&D-based and IP-based enterprises.

And the shrewd master here chastises the wicked servant for putting his lone talent in a hole in the ground instead of at least depositing it and earning interest.

Economic incentive for undertaking invention or creation is the same as the drive behind the master entrusting talents to his servants in this parable. He reasonably sought a return on his investments. In many circumstances in life, we have an opportunity to add value, whether at work, at home, or in the classroom. Part of the added value is the economic value of what we contribute or create. The prospect of earning more or creating new wealth based on the value of a new invention frequently sparks someone to embark on a creative endeavor.

The need for adding to one's personal economic pie often underlies applying one's abilities in discovery and invention. In this way, the reasonable assumption that one will own what one makes drives the application of his abilities and resources to a creative process. Here we see ownership's and creativity's mutual reinforcement with a payday, and often ongoing remuneration, in the offing.

Personal or Expressive Incentives

Some incentives to create fall outside the realm of the practical or remunerative. Sometimes we create just for the fun of it. Sometimes we're inspired. Sometimes it's for a diversion or some expressive motivation.

Whatever one's creative talents or motivations, they usually give the person a degree of pleasure. Being creative may serve as an outlet for relaxation. Maybe the creative outlet is unlike the demands, skills, and daily stresses of the job. Expressive incentive for creativity may be readily available while on the job, such as doodling in the margins of a notebook or making a rough sketch on a pad during a long meeting. Curiosity may be the driver. Or an intangible vision, idea, or inspiration could come out of nowhere. Being creative in this way differs from mere entertainment, say playing games on your smartphone. That's consumption.

Success. Recognition. Accomplishment. Fulfillment. Altruism. Wonderment. Discovery of new knowledge. Such nonremunerative motives may drive creative endeavor. While the inventor Nikola Tesla was motivated in part by the need to fund an upscale lifestyle, a combination of personal incentives and benefits also moved him.

When Tesla and inventor-industrialist George Westinghouse successfully connected their Niagara hydroelectric power station to the city of Buffalo and illuminated artificial light there, Tesla spoke at Buffalo's celebratory dinner on January 12, 1897, at the Ellicott Square Building. His remarks reflect noneconomic incentives behind his inventing. There Tesla commends to his audience the "spirit which makes men in all stages and position work, not as much for any material benefit or compensation, although reason may dictate this also, but for the sake of success, for the pleasure there is in achieving it and for the good they might do thereby to their fellow men."[9] Tesla acknowledges the economic incentive for work. But he continues by highlighting noneconomic incentives. He names success, pleasure in achievement, and the good of humanity

Similarly, Tuskegee Institute inventor George Washington Carver, who spent his career researching and developing a wide range of uses of peanuts and peanut by-products and crop rotation methods, grounded his calling to create and discover upon his walk with Christ. For him, invention was an extension of his Christian faith and his daily walk with God. "Only alone [in the laboratory] can I draw close enough to God to discover his secrets," Carver said while locking his lab door.[10] In addition to love of and communing with the Creator God, he also was motivated in part to express love for his neighbor through this occupational calling. Carver observed, "My purpose alone must be God's purpose—to increase the welfare and happiness of his people."[11]

Tying these types of incentives for creative activity to ownership, even something impromptu like sketching or doodling, produces something tangible. The tangible item belongs to its creator. This may seem obvious when the motivation is personal or expressive. However, stating this truth is important because the output of personal or expressive incentive may have intrinsic value. It would be out of step with Scripture's counsel to claim otherwise. For example, how valuable would a preliminary sketch by Michelangelo, Da Vinci, or Rembrandt be simply because of who made it?

If you go to your kitchen and make a batch of cookies, the cookies belong to you, the baker. If you spend an evening drawing or painting a picture, crocheting, making something of macrame, building a shadow box to display a collection, or fashioning a wreath for your front door, for instance, the creative output in each instance constitutes something you made. You bought or gathered the materials, and you put the effort into making the cookies, drawing, painting, crochet, macrame, shadow box, or wreath.

Whether your creation is music, a painting, an invention, or flowers or tomatoes from your garden, it's yours, yours to do with as you please. This illustrates how creativity and ownership are essentially mutually reinforcing, even when the principal reason for creating is merely personal or expressive.

● ● ●

To conclude our discussion of the mutual reinforcement of creativity and ownership, Susan M. Armstrong, a highly talented engineer at Qualcomm, the semiconductor innovator, helps weave it together. Armstrong's nascent creative interests took her into a highly technical field in which her creativity takes on a whole new form and brings a greater degree of fulfillment and flourishing. Her career has put Armstrong at the cutting edge of microchip functionality. ". . . I am an engineer; I like to make things, and I attribute some of that, quite frankly, to the fact that I'm crafty," she says. "I love to work with my hands, I love to make things."[12]

Hobbies in crafts factored into the educational and career paths this accomplished engineer has taken. "To me, computer science was you're trying to get this program to work correctly and you're trying to understand all the interactions with other systems and such," she says. "And to me it's just sewing on a big scale, but it was also puzzles and such. And when you come out, you come [to] these projects that you do in school or you do in a career setting, and you realize you've actually made something that is a benefit to a large number of people, not just for your own joy." It's worth noting how Armstrong honed her abilities of problem-solving and creativity in crafts and then in computer programming. "Problem-solving is pretty much why you invent," she adds. These endeavors yield joy from solving problems—first on a small scale and later on a large stage.

"And then from that to invention, I can create these unique inventions and I can share them with the world," Armstrong explains. Then naturally flow the ownership and economic elements. "At Qualcomm, we do a lot of work in, of course, in patents and such, [which] to me it's just this phenomenal way to be able to legally share with the world all these wonderful creations that you've come up with."

At Armstrong's employer, engineers constitute a large share of Qualcomm's workforce. Each one has his or her own story of coming to and learning problem-solving as part of the course taken to engineering. For the Fortune 500 company providing these thousands of engineers a creative outlet, the mutual reinforcement of creativity and ownership in these individuals explains part of its continued success as a global wireless technology standard setter and technological pioneer. The vital, mutually reinforcing combination manifests for both individuals and teams of individuals.

Chapter 11

Benefits from Owning What You Create

"A doctor can have a huge influence on your condition, but the human condition over time is improved by adventurers and entrepreneurs who invest in new ideas, and I'm living vicariously through those people. And when people ask me, will the next generation be worse off or better off than the generation we're in right now, I say it's almost certainly going to be better off. But not because of politicians. It's going to be because of the inventors and entrepreneurs and disruptors."
—U.S. Rep. Thomas Massie[1]

We have seen how practical, economic, and expressive or personal incentives may prompt someone to create or invent, with perhaps more than one kind of incentive driving a particular creative endeavor. Understanding the differences in motivation gives us a firmer biblical grasp on these aspects of the God-given human qualities of creativity and ownership. We now consider creativity-ownership's benefits.

The mutual reinforcement of ownership and creativity not only incentivizes creative endeavors, it also yields benefits. Specific benefits from this powerful combination include owning and controlling what you create, economic benefits derived from the monetary value of the creation, and the benefits of human flourishing. Of course, practical usage of one's creation counts as a benefit of the creativity-ownership combination. This we discuss at some length in other contexts. Thus, we won't cover the benefit of practical usage here.

Ownership and Control

A creative idea may be conceived only as something to do for fun. Or for using up some extra paint, glue, cloth, or other supplies. Or creative inspiration may be based on entrepreneurial aspirations, the prospects of inventing something around which you can build a business. Perhaps a part-time musician wants to give up the day job and be a full-time songwriter. A computer coder may aim to develop digital apps or computer games to license to a large firm and thereby fund more app or game creation. In all these instances, there is no doubt the creative endeavor is each individual's own idea, his or her own conception. If you are the creator, it is your time, energy, and effort invested in the endeavor. No matter how simple, complex, fancy, pretty, plain, valuable or common the creation, there is no doubt it is your creation, the fruit of your labor.

Nor is there any doubt what you created—the product, the artwork, the result—belongs to you. You invested your own money, your own resources, your own self into the creative endeavor. Therefore, what you made is a product of your intellectual and physical effort and is your property. As your property, you have sole discretion over what to do with the property. You may use it, save it, give it as a gift, throw it away, or adapt it to some other use or creative effort.

We plainly see both the creation and ownership of it are natural benefits for the creator. The creator-owner controls the property he or she has created.

Sometimes, one's creative output is developed as part of one's job. This is called a ***work made for hire***. Employed engineers and scientists often assign the inventions and intellectual property they create on the job to the company or the institution as a condition of employment. But unless you have assigned your IP rights under a contract, what you make belongs to you, free and clear.

Ownership and control over one's newly created property empowers a creator to decide whether or not to use the creative product himself or herself. Control also entails deciding when, how, and on what terms and conditions the creation may be used. For example, someone may buy a DVD or Blu-Ray copy of a motion picture, whose copyright a movie studio owns. The copyright holder typically specifies the copy may only be used for private viewing. He may not show the film in a theater and sell tickets to viewers. Control also includes the owner's determining for what if any compensation, for how long, and by whom the product may be used by another. With patented and copyrighted inventions and creations, ownership and control are secured through exclusive legal rights.

Control over one's own creative property amounts to the owner's having say-so over the creation, such as keeping or selling it. Burying it in the yard or storing it in an attic. Not doing anything with the creation or putting it to use right away. Giving it away or renting it out. Keeping it intact or ripping it up. Regarding it a prototype or a step along the way to further improvement.

If the creator-owner wants to commercialize the invention or creation, certain options come into play: licensing the invention to a single manufacturer or to multiple firms. Marketing it through third-party distributors, such as retailers and wholesalers, or selling it directly to consumers. There may be considerations about different people or companies having exclusive rights in specific geographic territories.

All these are aspects of the fruits of one's creative labor. And all this stems from God's design. Astute thinkers such as political philosopher John Locke and economist Adam Smith deserve some credit for expounding these biblically based truths in their fields of thought.

Of course, a creator may choose different uses or applications of his or her abilities and creativity in different circumstances. This too is an aspect of ownership and control. Such choices remain the creator's discretion. A famous artist may share his or her talent freely or be paid handsomely for performing, depending on the circumstances. The creative ability belongs to the artist, who alone gets to decide whether to be remunerated for sharing his or her artistry.

In early 2021, the world-famous cellist Yo-Yo Ma took the precaution of waiting a few moments after receiving his COVID-19 vaccination. Yo-Yo Ma commands significant fees for playing his cello in performances around the world. He is one of the greatest musicians of our time and is justifiably well compensated for sharing his talent.

Yo-Yo Ma brought his cello into the vaccination site because his instrument is too valuable to leave in the car. In the waiting area, he took his cello from its case as a few fortunate vaccine recipients sat nearby.

Yo-Yo Ma, whose talent bears high economic value, played Schubert's "Ave Maria" and other selections then and there, gratis, for the enjoyment of those at the vaccination clinic. No charge. No tickets required. Just a generous gesture by one of the best musicians alive.[2] Sharing one's talents with others, generosity associated with creativity, pays the giver—the owner—intangible dividends. These instances also are valuable fruits of one's creative labor. And such generous gestures carry no obligation.

Another example comes from Volvo: the three-point seatbelt invention.[3] An engineer with the Volvo automobile company invented an improved type of seat belt. He added a second belt, crossing from shoulder to lap, to the lap belt. The inventor, who as an employee had assigned any IP on his inventions to his employer, filed for a U.S. patent in 1959, and the patent issued in 1962. Volvo begun installing the new safety belt in 1959. The company shared the new seat belt design with other automakers. Opting to do so voluntarily, Volvo did not enforce patent exclusivity on this invention, in the interest of removing a barrier from other automakers' rapid adoption of the new design in the interest of safety.

Similarly, the Human Genome Project, which sequenced the entirety of human DNA, required participants to agree not to patent the data. The goal was immediate public access to the genetic sequencing data, as the effort involved other nations and public money to map what many regarded as a public good.

Meanwhile, a private company, Celera, recognized the commercial value of significant levels of genetic sequencing of the human genome. Those data could be protected by patents and then made available for a subscription or licensing fee. Ultimately, the public and private parties reached an accord and announced on June 26, 2000, the milestones of their respective human genome efforts.[4]

As a matter of ownership and control, both approaches and courses of action—voluntarily sharing one's work or pursuing commercial return from the creative fruits of one's labor—are legitimate. Indeed, the benefits of ownership and creation range from exclusive usage of one's own creative output to willingly relinquishing control over it. What someone is the original creator of belongs to her. And a creator is sovereign over his or her property.

This range of ownership and control is biblically permissible. Assigning control over your inventions' IP to your employer as a term of employment is merely trading your creative output for a paycheck and a job, and as inventor you likely receive a share of the royalties. This counts as enjoyment of the fruits of your labor. There is nothing necessarily more virtuous about giving away one's creative output for free, declining to obtain the intellectual property or fully exercising its IP protection. For no commercial enterprise, from startup to a large corporation, can stay in business and continue to provide jobs, contribute to the economy, or underwrite the discovery and the research and development (R&D) of innovative new products or goods, if it isn't profitable and remunerating its investors.

The fact is, secure ownership and control incentivize creativity. They are part and parcel of the inherent rights to the fruits of one's labor, having one's own vine and fig tree, and enjoying the shalom associated with property ownership, including of one's creative output. These are benefits and this is how mutual reinforcement works.

Economic Value

Another type of benefit produced by the mutual reinforcement of creativity and ownership is the prospect of economic value associated with a creation or an invention. Some creative products may be intrinsically valuable. Other creations become valuable because they prove to be technologically superior to alternatives, including prior solutions. Economic value derives from any of a range of factors: quality, scarcity, essentiality, ease of access, or the maker's reputation, for example. Uniquely painted ceramic dishes that a local craftsman makes to sell at craft fairs may hold a degree of value, but not nearly the same as Qing Dynasty porcelain or Wedgwood china, which gain value from their quality and scarcity. Some state-of-the-art inventions become the industry standard, which makes them essential and, therefore, quite valuable.

Property that holds intrinsic value makes it of great worth. Think of precious jewels, such as diamonds and rubies, or precious metals, such as gold and silver. "Rare earth" minerals critical for industrial purposes, such as the lithium used in electric vehicle and other batteries, are intrinsically valuable. Fuel sources, such as wood, oil, gas, electricity, and coal, are also valuable due to their importance to ensuring human survival and modern existence. Obviously, these are God-made creations humans extract and put to use.

Manmade creations having intrinsic economic value include the works of famous artists such as Rembrandt and Van Gogh. These masters aren't painting any more pictures, so this fact increases the value of their existing artwork. The ownership of these classic works of art has passed on from their creators to new owners, controllers of rights, or custodians of the masterpieces, such as museums.

Previously, we discussed the economic incentives for initiating creative effort. Picking up that thread, we now tie it to the benefit, the economic value derived from the mutual reinforcement of creativity and ownership. There may be a difference between a creator making a unique piece of expressive or useful art and selling it and producing something with commercial value that can be replicated. The first type of economic exchange and strategy for extracting economic value from a work resembles the painter who produces a portrait for a patron then paints a landscape and sells it to whomever wants it at the best price and on and on. This approach moves from one project to the next, some commissioned, others on speculation.

Another model holds certain advantages lending to a creative endeavor's potential to expand its economic value. This approach is scalable. For example, an artist may reproduce a painting as prints, selling multiple copies of the unique image. This model affords the owner more options, such as lining up different geographic markets; customizing terms and conditions for use, rights to make or sell, and other specifications to fit varying circumstances; and different price points depending on location, local markets, volume, or other factors. A commonly followed option is where an inventor or creator makes a creation the creator-owner can launch a business around.

Thus, for some, economic value aligns more closely with what a unique creation will sell for. For others, economic value is tied to entrepreneurial opportunity. Still others seek economic value from the risk-and-reward rate of return due to the creation's commercial value or high market demand.

For example, a songwriter may have composed a song that suits a superstar performer. Striking a deal where that song is sung by that performer may bring in outsized royalties for performances, both live and from recordings or downloaded purchases. That income may make it possible for the breakout songwriter to begin songwriting on a full-time basis. This is a form of specialization, doing what you are good at and partnering with others whose abilities complement your creative talents on the commercial side of the equation.

Many inventors have derived and increased economic value from their inventions by licensing them to others capable of product and market development, manufacturing, and the implementation or business side of taking an invention to market. It isn't every inventor who also has a head for business. Licensing allows someone skilled at discovery and invention to stick with doing what he or she does best.

* * *

An example of commercializing one's inventions involves an inventor-entrepreneur. This Clemson University Ph.D. bioengineer did a stint working in Clemson's technology transfer office helping to put university inventions on the path to commercialization. Matthew Gevaert, cofounder, board member, and former CEO of KIYATEC of Greenville, South Carolina, is transferring his university research discovery, follow-on inventions, and improvements into commercial use.

Founded in 2005, KIYATEC uses its Ex Vivo 3D Cell Culture technology or EV3D, and its 3D-Predict platform to yield more accurate predictions of a cancer patient's response to certain oncology medicines. It has been applied in connection with several cancers, including ovarian cancer, breast cancer, and glioblastoma. The technology uses a three-dimensional (3D) cell culture that contains a sample taken from a patient's cancerous tumor. The combination of 3D, which better models solid tumors in the human body than does 2D, and the sample from an individual patient, making the tests personalized, has shown promising results.

KIYATEC's technology in a 2020 study at a Buffalo, New York, cancer center doubled the time between gliomas recurring from the average four months to 7.9 months.[5] Another study yielded accurate predictions of a drug's working in an ovarian cancer patient. It predicted the correct response thirty-nine out of forty-four times, or 89 percent. For a subset of twenty-nine of these patients, EV3D predicted a drug's efficacy and tolerance with 100 percent accuracy.[6] Such strong predictive evidence has helped KIYATEC attract backing from such venture capital (VC) investors as Boston VC firm Seae Ventures, and from the National Cancer Institute.[7] KIYATEC's progress continues, substantiated by clinical evidence, attracting additional investment, such as from two top brain tumor patient groups' venture philanthropy funds in 2023.[8]

● ● ●

Some inventions lay the foundation of a new kind of technology, such as the electrical generation and transmission system George Westinghouse and Nikola Tesla invented, collaborated on, and deployed at the 1893 Chicago World's Fair and at Niagara Falls. It had tremendous economic value. And the foundational inventions in each generation of wireless technology have proven especially valuable. They allow technology implementers, such as smartphone or auto makers, to use these critical inventions and build on the new system.

For example, fourth-generation, or 4G, wireless technological standards coupled high-speed streaming media capability to 3G's mobile broadband, digital voice, and data connectivity. The 4G foundational, standardized technology enhanced smartphone applications, or apps. Thus, thanks to the years of work on the 4G foundation by trail-blazing companies such as Qualcomm and InterDigital, implementers such as Apple and Google could develop devices such as their iPhone and Android smartphones. The 4G system improved connectivity markedly by reducing dropped calls and paved the way for the app economy. App and device developers could integrate the new wireless features and capabilities by licensing Qualcomm's patents for GPS (global positioning system) and app stores, for example.[9] Now, 5G has come to fruition, with greater data transmission capacity and speed. This makes the Internet of Things, or widespread connectivity, possible.

The point is the economic value of a creative or inventive work of a human being varies, depending on a number of factors. As seen, such value may come about in a variety of ways. And economic value depends in part on the nature of the creation's inherent value, on the economic worth those active in a competitive private market assign to an item, and on how much value may eventually be added to the invention or creation up and down the value chain.[10]

Those creations holding greater value are usually worth protecting as intellectual property, which we'll discuss in more detail in coming chapters. Whatever the value in a specific case, economic value and the commercial importance of one's creation are likely to be among the benefits derived from creative effort.

Human Flourishing

Corresponding to personal incentives for exerting creative effort, human flourishing counts as a benefit of ownership. It stems from the mutual reinforcement of one's creative abilities and ownership of what that person makes. As we've said, to flourish means to thrive—to excel, to succeed, to prosper. While the term often applies to financial prosperity, here we are talking about flourishing more in the sense of a plant growing luxuriantly. The way someone applies his or her skills, knowledge, and ability, along with his ownership stake and benefits from what his talents create, are the soil, seed, sun, and water that cause a bountiful harvest. In creating, we find fullness of heart.

Author and former editor at *Christianity Today* Andy Crouch regards human "flourishing as fullness of being—the 'life, and that abundantly' that Jesus spoke of. Flourishing refers to what you find when all the latent potential and possibility within any created thing or person are fully expressed . . . the transition from nature to culture as a move from 'good' to 'very good.'"[11]

Medical device inventor Cephus Simmons cites a form of human flourishing proceeding from laboring through the invention, patenting, product development, and commercialization processes. His accomplishments resulted in "having a product on the market, and it's benefiting and helping children and patients of all ages; it is a glorious thing to be able to look back and see how God gave me something that's really benefiting the community."[12]

Psalm 103 puts human flourishing in broader context. This psalm calls us to bless the Heavenly Father, who is the creator and sustainer of the universe, because he benefits his children with forgiveness of sins, healing, redemption, love, mercy, good things, compassion, and "he remembers that we are dust" (v. 14). It acknowledges humanity's temporal limitations, at the same time acknowledging God has made his children for flourishing while on earth. "As for man, his days are like grass; he flourishes like a flower of the field" (v. 15). This underscores human creativity, its role in God's plan, and how the creativity God gives us brings flourishing from our creating.

* * *

Thinking of human flourishing or thriving reminds me of a scene from the movie *Chariots of Fire*.[13] The athletic talent of Scottish rugby-player-turned-runner Eric Liddell set him up to run for Great Britain in the 1924 Olympics. The Liddell siblings born to missionary parents in China were intent on returning to China to continue in Christian ministry. But the demands of Eric's athletic training and competition concern his sister, Jennie. She worries her brother's commitment to their missionary calling may be slipping. She ardently expresses this fear to Eric.

Eric and Jennie walk in the hills. In their peripatetic conversation, Eric reassures her the China mission remains his plan, but the Lord also has called him to run in the Olympics. Eric turns to Jennie and says, "I believe that God made me for a purpose. For China. But he also made me fast. And when I run, I feel his pleasure."

Eric Liddell had God-given ability as a runner and the opportunity to use that ability representing his country—and the Lord, as it turned out, when he refused to run in a qualifying heat on the Sabbath—on the international stage. God made clear to Liddell he had both short-term and long-term calls upon Eric's life. Amidst the efforts, Liddell found fulfillment in using his talent for God's glory in the circumstances the Lord opened for him. In the moment, Liddell's faithfulness and pursuit of excellence in the calling by application of his talents gave Eric a feeling of pleasure, joy, and peace under the Lord's smile. That's what's meant here by human flourishing.

• • •

An example of human flourishing made an impression on me early on. When I was growing up, our home became a mass-production operation as Christmas neared. My mother, who was artistic, came into her own in this annual manifestation of creativity.

Mama would bake, bake, bake. Cookies! Gumdrop cookies. Toll House cookies. Fruitcake cookies. Massive amounts of cookies, with an appreciable share frozen for enjoyment over the warmer months.

More! Divinity. Orange-coconut ambrosia. Pecan and other pies. Fruitcakes. Cheese biscuits. And even use of the inedible: hollowed walnut shells inside which she'd put a shiny penny or other keepsake and glue the halves of the shell back together, each becoming a beloved treasure of nieces, nephews, and other children.

Mama thoroughly enjoyed making these goodies. She also enjoyed sharing these treats. They'd be delivered as gifts to friends, to relatives, to Daddy's coworkers. They were served to guests in our home. They enriched the Christmas season for all of us. The whole creative and distribution process gave my mother tremendous joy. (Not to be outdone, Daddy made the very best peanut brittle I've ever tasted, and according to everyone who ever tasted it. He made lots and lots and gave peanut brittle as gifts or shared it with visitors in our home. To be sure, this was not a competition between them, but complementary creativity.) Truly, this yearly effort by my parents was a highlight of the Christmas season.

My parents' joy came from the creative exercise. It came from the hands-on process of making something from various ingredients, that is, raw materials. It came from satisfaction derived with each finished batch. It came from the completion of the entire production. It came from sharing with others what they made—delivering these homemade offerings, seeing the smiles, knowing the recipients appreciated these tasty tokens. It came from being creative and sensing God's smile on their use of the abilities and interests he gave them.

This seasonal creative endeavor constituted a family tradition, a practical display of love for neighbor, sharing the blessings God gave my parents. My mother was a creator. She painted with oils and watercolor on canvas, ceramics, and other media. She drew with pen and ink, made and designed dolls and various crafts. She sometimes played piano. But in their yearly Yuletide labor of love, my mother and father honored the Lord, used the gifts he bestowed, their creative abilities, and enriched many people's Christmases while honoring the newborn King in a manger in Bethlehem. Observing this festive creativity impressed upon me something I came to understand as human flourishing.

● ● ●

Whether someone is driven to create by practical incentive, economic incentive, or expressive or personal incentive, creative effort produces benefits, as we've been examining. We see how the benefits generally correspond with the several types of incentives spurred by the creativity-ownership combination. That old fuel of interest added to the fire of genius produces something out of our own creativity, and we get something out of it, tangible and intangible, as God designed.

Usually, creative product has a purpose. Its purpose is wrapped up in what led us to pursue the artistic, mechanical, or scientific effort. Thus, we actually seek some benefit or benefits from our creative output. We've seen these benefits range from owning our creation and the freedom to use, make, sell, give away, or license the items we've come up with. Or more personally or expressively, we garner satisfaction, fulfillment, joy, pride, self-worth, accomplishment or wonderment from our creative output. That is, our ideas and efforts yield certain fruit, thanks to divinely endowed creativity and ownership.

The benefits, whatever their flavors, are part of the fruit of our labor. And to the one who made the beneficial creation rightly belongs its fruits. No creation of human hands is by nature community property. It's preordained to belong to the person who made it, just as God's creatures belong to him.

Chapter 12
Property Rights

"In a word, as a man is said to have a right to his property, he may be equally said to have a property in his rights."
— James Madison

In his tour de force *The Wealth and Poverty of Nations*, history and economics professor David S. Landes explores and explains why some countries are rich and others poor. Among the reasons for vast differences in the economic well-being of various cultures, Landes writes, are property rights.

> Linked to the opposition between Greek democracy and oriental despotism was that between private property and ruler-owns-all. Indeed, that was the salient characteristic of despotism, that the ruler, who was viewed as a god or as partaking of the divine, thus different from and far above his subjects, could do as he pleased with their lives and things, which they held at his pleasure. . . .
>
> Today, of course, we recognize that such contingency of ownership stifles enterprise and stunts development; for why should anyone invest capital or labor in the creation or acquisition of wealth that he may not be allowed to keep? In the words of [British statesman] Edmund Burke, "a law against property is a law against industry."[1]

Landes describes how European nations, Britain in particular, put into place the building blocks that gave rise to the Industrial Revolution in the nineteenth century. Among these components were secure private property rights.[2]

The inherent right of private ownership entails the right and freedom to own property and to enjoy the fruits of one's labor. Private property rights extend more broadly than to land, financial assets, and personal possessions. The right to own property includes owning the fruits of one's creative labor, such as inventions, writings, paintings, and discoveries. The concept of property rights will help in understanding ownership in deeper ways. Also, a discussion of property rights will lay a foundation for examination of intellectual property rights.

A Return to Property Rights in the Bible

We begin our look at property rights with a return to their basis in Scripture. As explored in chapter 9, the right of ownership by human beings is God-given. We saw how the Bible bears out this fact.

The Ten Commandments require us to love God and our neighbor through obedience to God's law (they, of course, prove we cannot obey perfectly and thus need a righteous Savior, Jesus). In the Decalogue, the Lord commands our respecting what belongs to him and to other people. The first four commandments (Deut. 5:7–15) address respect for what belongs to our Maker. The next six (vv. 16–21) deal with respect for property that does not belong to us, but to another person. In the Sermon on the Mount (Matt. 5:17–32), Jesus illustrated how the spirit of God's commandments exceeds the letter of his law. Each commandment addresses a different aspect of private property.

The Eighth Commandment against stealing is a primary piece of evidence for individual property rights. It says, "You shall not steal." This law covers all forms of theft, fraud, and taking of another's property, whatever the form of the property. Taking something that doesn't belong to you displays ingratitude to God the provider and contempt for his provision for your needs. This is why the answer to Westminster Larger Catechism question 142 names "idleness, prodigality, [and] wasteful gaming" among the things that constitute stealing in violation of the Eighth Commandment.[3] Laziness is sinful failure to apply one's talents, abilities, and intellect and to assume personal responsibility, better oneself, and improve one's circumstances.

The Tenth Commandment prohibits "covet[ing] your neighbor's wife [or husband] . . . your neighbor's house or land, his male or female servant, his ox or donkey, or anything that belongs to your neighbor" (NIV). Coveting is wanting something that is another person's. Not just a car *like* my neighbor drives, but *that* car. Not just a house on a lot situated similarly on the street, hilltop, beach, neighborhood, etc. where my neighbor lives, but that house and land. Dwelling on and feeding such evil desires directly breaks God's command. The more such thoughts are fostered and indulged not only deepens this blatant, inward sin, it also increases the likelihood of acting on the iniquitous desire.

The Fifth through Seventh and Ninth Commandments also speak to something one's neighbor owns and is due. The Fifth Commandment says, "Honor your father and your mother, as the Lord your God has commanded you, so that you may live long and that it may go well with you in the land the Lord your God is giving you" (NIV). The Creator of our first parents, Adam and Eve, made them to be human parents, who gave life to children, who in turn owe honor, respect, and obedience to their earthly parents and their heavenly Creator. Though our first parents failed, the order of creation, with parents holding authority over their household, survives the Fall. The model extends to other authorities, such as ecclesiastical and civil.[4]

The Sixth Commandment, "You shall not murder," commands respect for and calls for the protection of human life. As we shall see, human life itself is a form of private property. The Seventh Commandment, "You shall not commit adultery," prohibits trespassing on someone else's home turf, figuratively speaking. Adultery encroaches on the private "property" of a marital covenant, intruding on intimacy a husband and wife jointly possess.[5] The Ninth Commandment, "You shall not give false testimony against your neighbor" (NIV), demands truthfulness in our interactions with or about our neighbor. The immediate context is false witness in court proceedings (see Deut. 19:15–21 for the penalty for false witness). The commandment's larger sense prohibits spreading falsehoods about one's neighbor, besmirching someone's reputation, or otherwise trafficking in misinformation about a person.[6]

• • •

Some have considered Acts 2:42–47 and 4:32–5:11, where Christian believers together in Jerusalem "had everything in common," selling their possessions and giving "to anyone who had need" (NIV), as a New Testament model for communal ownership. With resurgent interest in socialism, we do well to clear up what Acts communicates, and whether it really teaches socialism.

Is socialism compatible with Christianity? Does it square with God-given gifts of creativity and ownership rights? The situation recounted in Acts 2 and 4 is a surge of thousands of new Jesus-followers where the disciples congregated. A temporary influx had occurred in Jerusalem, where thousands of people came for the Feast of Weeks. The Lord used this occasion dramatically to send the Holy Spirit (Acts 1:4–8; 2:1–6, 38–41), the Comforter whom Jesus promised would come after Christ ascended to his glory (Luke 24:46–49, 52; John 14:16–17).

Acts 2:42–47 and 4:32–5:11 report periodic acts of generosity as needs arose under the circumstances during Pentecost. There is no once-for-all divestiture of all property, no church- or state-coerced taking of property or forcible redistribution of wealth. Material equality among believers and forcing believers to give up their private property rights are absent. Some among the early church during that Pentecost in Jerusalem shared voluntarily, without coercion. Pastor and author John Piper notes, "'Thou shalt not steal' makes no sense where no one has a right to keep what is his."[7]

Similarly, Harold Lindsell writes that socialism's goals include "the abolition of private property which is essential to economic sufficiency and to meet the economic needs of men."[8] Lindsell explains socialism's foundation: "In the broadest sense socialism is opposed to private ownership of the means of production. . . . Marxism and utopian socialism share the same ultimate objective, i.e., the organization of society in such a way that the workers own, control, and operate everything cooperatively."[9] He points out how socialism in practice never works out to be egalitarian or cooperative. Lindsell notes, "any form of socialism . . . is antithetical to free enterprise—which alone can be validated by the Judeo-Christian tradition. . . ."[10]

Upon examination, Acts 2 does not constitute a socialistic manifesto or repeal the Eighth and Tenth Commandments. The whole counsel of Scripture teaches private property rights along with individuals' voluntary acts of generosity and a lighter grip on possessions (e.g., Matt. 5:40; Heb. 10:34).

• • •

This look at the Decalogue and Acts highlights their property rights implications. The Lord gave these commandments, each attached to a certain type of property. We now turn to the property rights Landes credits with helping make the difference in some countries' wealth creation and their absence as causing others' economic stagnation.

John Locke, Property Rights, and America's Founding

Property rights in the West have roots in the Ten Commandments and other biblical teachings, the Greek philosopher Aristotle, and other sources, but perhaps foremost is seventeenth-century English political philosopher John Locke.[11] Locke articulates the individual right to "private property that is based on the natural right of one's ownership of one's own labor, and the right to nature's common property to the extent that one's labor can utilize it."[12] The notion of a law of nature, universally known to human beings and governing their relations among one another, aligns with the principles of the Decalogue, particularly the Eighth and Tenth Commandments relating to private property rights. Locke echoes the biblical teaching that the laws of nature come from a Lawgiver, who reveals himself to humans. "For since the creation of the world God's invisible qualities—his eternal power and divine nature—have been clearly seen, being understood from what has been made, so that men are without excuse" (Rom. 1:20 NIV).

Locke, in his *Second Treatise of Government*, connects a property right in oneself and thus to one's labor. He states, "Every man has a property in his own person: this nobody has any right to but himself. The labour of his body, and the work of his hands, we may say, are properly his."[13] This right of ownership of oneself and of one's own labor and the fruits of that labor as private property accord with and are corollaries of the Creation Mandate God gave Adam and Eve before the Fall (Gen. 1:28–31).

Of note, given our focus on human creativity and ownership, Locke applies this property right in oneself and one's labor to the fruits of ingenuity. "Man, by being master of himself, and proprietor of his own person, and the actions or labour of it, had still in himself the great foundation of property; and that, which made up the great part of what he applied to the support or comfort of his being, *when invention and arts had improved the conveniences of life, was perfectly his own, and did not belong in common to others*"[14] (emphasis added).

Locke includes the right to acquire property by appropriating common property found in a state of nature. He ties this to developing something from what is taken out of its natural state. "Whatsoever then he removes out of the state that nature hath provided, and left it in, he hath mixed his labour with, and joined to it something that is his own, and thereby makes it his property."[15] In other words, if you make something of unowned material, then you own it—or, we might add, if done working for someone else, you are due compensation for your labor.

Locke reasons that such developed natural resources benefited both the person exercising the right to property ownership through his or her labor and humankind broadly. "He who appropriates land to himself by his labour, does not lessen, but increase the common stock of mankind: for the provisions serving to the support of human life, produced by one acre of inclosed and cultivated land, are (to speak much within compass) ten times more than those which are yielded by an acre of land of an equal richness lying waste in common."[16]

In other words, investing one's labor into improving raw materials is the source of creating wealth. Individual people vary in terms of talent and abilities, as well as station in life. Property rights in one's labor mitigates such variation in faculties, providing the means to improve one's circumstances. Working hard turns time and effort into earnings. Earnings—a fruit of one's labor—if saved, enable an earner to acquire more property.[17]

Locke places property rights, centrally the property interest in one's labor, in the frame of the Creation Mandate and Common Grace. "God, when he gave the world in common to all mankind, commanded man also to labour, and the penury of his condition required it of him. God and his reason commanded him to subdue the earth, i.e. improve it for the benefit of life, and therein lay out something upon it that was his own, his labour. He that in obedience to this command of God, subdued, tilled and sowed any part of it, thereby annexed to it something that was his property, which another had no title to, nor could without injury take from him."[18]

John Locke lived in the 1600s, a period when the West was coming into its own as a civilization. Europe, North America, and European-settled nations—Christendom—benefited from profound, culture-shaping experiences: The classical influences on law, ethics, and government; the Renaissance; the Protestant Reformation, and the Bible's accessibility;[19] the Enlightenment; legal developments, such as the Magna Carta, natural law, and common law; and representative government, in commonwealth, republican, and democratic forms.[20] These cultural features and thought—individual rights, separation of civil and religious authority, spreading literacy and literature, the rule of law, for instance—over time influenced Locke and Western territories. In turn, Locke's influence on Western property rights doctrine was significant, including in the American colonies.

* * *

The Declaration of Independence succinctly summarizes essential components of Western thinking on individual rights, law, and government:

> We hold these Truths to be self-evident, that all Men are created equal, that they are endowed by their Creator with certain unalienable Rights, that among these are Life, Liberty, and the Pursuit of Happiness—That to secure these Rights, Governments are instituted among Men, deriving their just Powers from the Consent of the Governed, that whenever any Form of Government becomes destructive of these Ends, it is the Right of the People to alter or to abolish it, and to institute new Government, laying its Foundation on such Principles, and organizing its Powers in such Form, as to them shall seem most likely to effect their Safety and Happiness.[21]

Additionally, the Declaration's first paragraph appeals to "the Laws of Nature and of Nature's God" as the basis for the American colonies' severing political ties with Britain. These passages contain terms appearing in Locke's writings. "Natural liberty." "Laws of nature." "Consent of the [governed] people."

The choice of the words "Life, Liberty, and the Pursuit of Happiness" for examples of "certain unalienable Rights" has generated discussion as to whether "pursuit of happiness" means "property." The short answer is yes, and more. The scholar Edward Erler observes, "The right to property was understood by the founders as the comprehensive right that included all other rights [e.g., equality under the law, private property rights, liberty of conscience]. Understood in this manner, the right to property was described in our most authoritative document as the 'pursuit of happiness,' which was considered not only a natural right but also a moral obligation."[22]

The Virginia Declaration of Rights, written by George Mason and adopted June 12, 1776, served as a reference for Thomas Jefferson's drafting of the Declaration of Independence that summer. It begins: "That all men are by nature equally free and independent and have certain inherent rights, of which, when they enter into a state of society, they cannot, by any compact, deprive or divest their posterity; namely, *the enjoyment of life and liberty*, with the *means of acquiring and possessing property*, and *pursuing and obtaining happiness and safety*" (emphasis added).[23]

This model for the Declaration of Independence echoes key Lockean principles, such as laws of nature that bestow life, liberty, equality, consent of the governed, and property. The Virginia Declaration also invokes the pursuit of happiness and safety—all found in Locke. Jefferson tightened and recast these ideas.

Following the Declaration of Independence, the American Revolution bought the colonies their liberty. When the first form of national government, the Articles of Confederation, proved inadequate, delegates to the 1787 Constitutional Convention developed, adopted, and put forth the U.S. Constitution for the states to ratify. The preamble to the Constitution explains the purpose and goals of this newly designed national government: "We the People of the United States, in Order to form a more perfect Union, establish Justice, insure domestic Tranquility, provide for the common defence, promote the general Welfare, and secure the Blessings of Liberty to ourselves and our Posterity, do ordain and establish this Constitution for the United States of America."[24]

Protecting citizens' life, liberty, private property, safety, and happiness all shine through in this preamble to the document ordaining America's republican government. The Founders, called Federalists, who supported replacing the Confederation government with the republic described in the Constitution, advocated for ratification by the states. The eighty-five essays of the *Federalist Papers* explained the reasons for the proposal and supported the proposed changes.

Recall, "the pursuit of happiness" broadly means various forms of property. Federalist 10, penned by James Madison, says "the first object of government" is to protect individuals' abilities "from which the rights of property originate" (that is, the means of work and enjoyment of their labors' fruits) and to "acquir[e] property."[25] This appeals to Locke's "chief end" for why people "[put] themselves under government . . . the preservation of their property."[26] Civil government's first duty is protecting property rights.

Individuals differ by a wide range of "faculties"—abilities, talents, opportunities, etc.—which often results in their forming factions. Madison observes, "The most common and durable source of factions has been the various and unequal distribution of property." Faction not only may prove divisive and disruptive in public life and private dealings. It risks impeding the very institution whose "first object" is to protect individual property rights, liberty, and life from fulfilling this central task—thus leaving property rights unsecured. Federalist 10 explains how the new Constitution's form of government is best suited to promote a sufficiently strong union and to diffuse and diminish such harmful effects of faction.

About a century after publication of Locke's *Treatises of Government*, James Madison—the Father of the Constitution—pithily captured the crux of Locke's property rights principles: "In a word, as a man is said to have a right to his property, he may be equally said to have a property in his rights."[27] Madison posits the gist of property rights emanating from the Reformation and the Lockean Enlightenment.[28]

Founding Father John Adams expresses the moral aspect of property rights, connecting property rights with God's law. Notably, Adams ties the sanctity of property to the Sixth and Eighth Commandments proscribing stealing and coveting, respectively. "The moment the idea is admitted into society, that property is not as sacred as the laws of God, and that there is not a force of law and public justice to protect it, anarchy and tyranny commence. If 'Thou shalt not covet,' and 'Thou shalt not steal,' were not commandments of Heaven, they must be made inviolable precepts in every society, before it can be civilized or made free."[29]

● ● ●

The American property rights tradition derives from the Bible, John Locke and other great minds of the Western world, and a providential group of Founders steeped in these elements of the Western tradition. Lockean property rights philosophy has suited European and North American nations, particularly the United States, because it stems from Judeo-Christian Scripture and the Reformation, as well as the classical and rational influences of ancient Greece and Rome and the Enlightenment.

Private property rights have thrived in America due to our Old World roots and our New World perspective. America differed from Europe in ways that meant private property was more front and center than in ancient lands where monarchs and stark class differences were prominent. Samuel Huntington puts it, "America was the land of freedom, equality, opportunity, the future . . ."[30]

The Fruits of Private Property

Our brief survey of property rights returns us to Landes's difference "between private property and ruler-owns-all" characterizing Western and Eastern civilizations, respectively. The difference has proven significant.

Nobel economist Edmund Phelps has extensively analyzed economic performance and national prosperity on a range of measures. The data, for example, show economic growth is slower in nations with "high SOE [state-owned enterprise] per unit of GDP [gross domestic product]."[31] State ownership of industrial and commercial companies characterizes socialist economics.

Phelps offers another possible explanation for socialism's poor economic performance: "It could be that high state ownership and low growth are both effects of a third influence—a heedlessness toward property rights or an outright antagonism toward private property . . ."[32] Phelps's findings confirm this correlation: "Where a country opposes private ownership of enterprises—where it is socialist minded—it suffers poor economic performance."[33] That is, denial of private property rights costs an economy or a state in lost prosperity and lost flourishing.

Property rights, coupled with ordered liberty, the rule of law, and limited government, provide a stimulative environment for thriving like a greenhouse for growing plants. "When property is secure, and cannot be taken away by violence, lawlessness, or arbitrary government decree, there is an incentive to earn more, save more, and invest in more opportunities for the future—which encourage enterprise and further economic activity," writes Matthew Spalding. "If you are guaranteed to reap what you sow, then there will be more people sowing and reaping, leading to economic growth. This holds true not just for individual entrepreneurs but for small businesses, companies, and large corporations as well. As a result of that activity being available to all, and of the protection of property extending to all, the overall level of wealth in the whole society will expand."[34]

President Ronald Reagan lauds the private ownership of property as a bulwark of opportunity and freedom. "I've long believed that one of the mainsprings of our own liberty has been the widespread ownership of property among our people and the expectation that anyone's child, even from the humblest of families, could grow up to own a business or a corporation," Reagan says. "Thomas Jefferson dreamed of a land of small farmers, of shop owners, and merchants. Abraham Lincoln signed into law the Homestead Act that ensured that the great western prairies of America would be the realm of independent, property-owning citizens—a mightier guarantee of freedom is difficult to imagine."[35]

This assessment of some key benefits property rights yield aptly brings home the realization of hopes and dreams that property rights make possible. Kirk, in *The Politics of Prudence*, summarized important benefits of property rights: "To be able to retain the fruits of one's labor; to be able to see one's work made permanent; to be able to bequeath one's property to one's posterity; to be able to rise from the natural condition of grinding poverty to the security of enduring accomplishment; to have something that is really one's own—these are advantages [of secure property rights that are] difficult to deny."[36]

* * *

As we have seen, rights of property ownership act as an enabler and an incentive to work, providing individuals the means to acquire property of various types. Put another way, secure, enforceable, legal property rights empower a person's labor, industriousness, and economic endeavor, which produces fruit, literal or figurative. Individuals naturally own the output or are free to exchange their labor for payment to compensate for their expended time, energy, and effort. If someone's creative output is developed as part of one's job—a work made for hire—that person is owed compensation for it—it's still fruit.

Another common description of property rights is "putting food on the table." This image implies an individual's ownership of the fruits his or her labor produces. Psalm 128:2a affirms this fact of property rights in the creativity-ownership combination: "You shall eat the fruit of the labor of your hands."

Chapter 13
Intellectual Property Basics

"... IP is useful in two distinct ways: First, it protects you from people copying your invention, essentially giving you a monopoly on your technology if you obtain broad patent coverage; second, it also represents equity. IP represents value ..."
— Robert Yonover, inventor[1]

"We protect intellectual property, the labors of the mind, ... as much a man's own, and as much the fruit of his honest industry, as the wheat he cultivates, or the flocks he rears."
— Justice Levi Woodbury in *Davoll v. Brown* (1845)[2]

Life's circumstances can present opportunities to create or innovate. We get ideas for solving a problem or something to act on. Our ideas, whether in the workplace, at home, or the classroom, may work and may potentially be valuable. A creative idea may come to be worth something commercially. Therefore, it may be worth protecting as intellectual property.

An idea itself isn't eligible for intellectual property (IP) protection. Ideas are "intangible, even ephemeral."[3] Ideas may launch or guide turning one's idea into something tangible and perhaps useful and valuable. "Ideas catalyze the birth of a book, invention, song, movie, or any other creation. And although ideas cannot be owned, the right to exclude others from producing the creations that flow from them can be—at least for a set time."[4] An idea is like a vision for an original design of a building. An idea isn't a new building or even a new building design; it's the mental conceptualization of such a design while IP is more like a building blueprint. IP protects the ownership rights of the tangible fruit of an idea.

The primary forms of IP are copyright, patent, trademark, and trade secret. Their common denominator is each secures legal rights in newly created property, that is, in property a human being has created. IP, whether a copyright, patent, or trademark, specifies the boundaries of the property through a process in which a government office plays a role registering the private property. It's similar to what transpires in land purchases: surveyors precisely measure and mark the metes and bounds of a plot of land. Then the register of deeds in the jurisdiction in which the land lies records the property information and the ownership transaction for public records.

Much like a registrar of deeds, the U.S. Patent and Trademark Office (PTO) examines patent applications and records an invention's particulars if a patent is issued on an invention. PTO also registers trademarks, assuming another firm has not already claimed the name or logo submitted. The U.S. Copyright Office serves this recordation function for new compositions or visual images. These government recordation services for IP are as important as other property records, both as a legal matter and as an impartial record of property holdings of a particular form of property. Trade secrets protect IP through private contract rather than public process.

Once obtained, IP promises the owners the exclusive right to make, use, sell, or license the IP. That is to say, IP secures the right to exclude others from making, using, selling, or licensing that invention, motion picture, discovery, book, photograph or other protected creation. As private property with registered title, IP may be sold or bought, assigned to another or kept or licensed—granting limited permission to another or others to use the IP in certain ways for a certain period of time on certain terms. IP rights carry legal consequences; the IP owner has the right to enforce his or her exclusive rights against IP infringers. Think of a copyright, patent, or trademark like a "no trespassing" sign to keep other people off your property or like a license plate on your car or pickup truck, which indicates private ownership of the vehicle and the ownership interest's recordation with state government.

In exchange for exclusive ownership of the newly created intellectual property for a limited time, patentees, copyright holders, and trademark owners must publicly share certain information about the invention, creation, or mark. The IP protected as a trade secret is just that: a closely guarded secret.

* * *

Inventor Robert Yonover and Ellie Crowe explain IP's value: ". . . IP is useful in two distinct ways: First, it protects you from people copying your invention [or other creation], essentially giving you a monopoly on your technology [or other creation] if you obtain broad patent [or another form of IP] coverage; second, it also represents equity."[5] IP represents value and can be sold, so you want to build up your IP to possibly sell or license."[6] I take issue with using "monopoly" for IP exclusivity, but we'll get to that later.

Another way of stating Yonover and Crowe's second point is IP is an intangible asset. *Black's Law Dictionary* defines such "intangible property" as "property as has no intrinsic or marketable value but is merely the representative or evidence of value."[7] IP acts as a deed or title to valuable property. IP's value lies not in the paper the patent is printed on; it resides in the invention a patent describes, the worth investors and licensees attribute to it, and the interest other players in the market show in the IP.

The phrase "build up your IP" raises two aspects to note. First, it's common to protect intellectual property interests in an ingenious creation with more than one patent. There may be a patent on a new machine (or device or compound), one on a newly developed material the device is made of, another on a process for manufacturing the material, and one on a separate process for making the machine. Second, a new product may be protected by a patent or several patents, while a formula or method of manufacture is retained as a trade secret, the owner's manual for the product is copyrighted, and the product's brand name is trademarked.

Before we examine the four main types of IP in turn, an illustration coined by IP attorney Lawrence J. Siskind differentiates the protection each IP form provides. If Thomas Edison wrote a book about the incandescence process, a copyright would protect his written composition and graphic images from being copied, but it wouldn't stop someone from using the inventive information to make electric light bulbs. A patent would protect against unauthorized use of Edison's invention described in his book until the patent term expired and the invention entered the public domain. A trademark would protect the name or other unique features Edison gave his incandescent light bulb. While other light bulbs could be marketed, they couldn't carry the same branding elements Edison used for marketing his bulbs. Finally, Edison could keep secret some aspects related to his brand-name light bulb or the process for making them, relying on trade secret protection.[8]

The Main Types of IP

Intellectual property comes in many forms, including inventions, brand names, discoveries, recordings of motion or sound, photographs, logos, designs, databases, software, processes, methods, and other types of created works. We will focus on the primary forms of intellectual property under which these other types of IP are legally protected: copyright, patent, trademark, and trade secret.

Copyright

Copyright protects an original work that's fixed in a tangible medium. That is, a created work must be on film, on videotape, in digital format, on canvas, paper, or other medium from which it may be communicated to other people. A copyright secures the exclusive right to print, publish, perform, or record certain literary, artistic, or musical material.

The owner exclusively has the right to use and distribute the copyrighted item. She or he also may authorize (e.g., license) others to use or sell the copyrighted work. Copyright covers, for example, writings, photographs, and musical or audiovisual recordings or compositions. Movies are protected by copyright, as are computer software programs (i.e., the code itself). Creative works typically bear the copyright symbol (©), though fixing an original expression in a tangible medium provides rights.

The origins of copyright had less to do with securing private property rights and more about government control. Gutenberg's printing press invention in the fifteenth century became a tool for spreading the Bible in common languages and literature containing the arguments of the Protestant Reformation. The British monarch in the mid-sixteenth century established a monopoly on printing in the kingdom. "By requiring that all published works be registered with the [crown's copyright holder]

Stationers' Company, the government made it easier to block dissemination of heretical writings."[9] In 1710, the British government enacted the Statute of Anne, allowing authors exclusive rights to publish their own original works and reducing publishers' monopoly power.[10]

The key terms in relation to copyright are "original work," "tangible medium," and "exclusive right." "Original work" means something a person independently created involving some level of thought.[11]

"Tangible medium" means that in which a sound, composition, image, or likeness is fixed. A camera's shutter captures an image on film, tape, or digitally. I'm typing the words of this chapter on my MacBook Pro laptop; my fingers typing on the keyboard fixes my thoughts in digitally preserved words (I'm ever grateful to my mother for encouraging me to take typing class in high school.). Under the 1976 Copyright Act, a creative work becomes fixed in a tangible medium and thus is protected "as soon as a work is recorded in some concrete way."[12]

"Exclusive right" under copyright is broad and general but has its limits under the law. First, copyright law provides for "fair use" of otherwise IP-protected creative works. Fair use doctrine attempts to balance creators' ownership interests with legitimate uses. This exception "allowed library photocopying . . . for purposes of scholarship, preservation, and interlibrary loan . . . [and] for purposes of criticism, commentary, news reporting, teaching, scholarship, and research."[13]

It grants people who don't hold a work's copyright a privilege "to use the copyrighted material in a reasonable manner without the owner's consent."[14] Whether unauthorized use constitutes fair use depends on four factors: (1) the use's purpose; (2) the nature of the material used; (3) how much of the entire work is used (less is fair, more becomes questionable); and (4) the use's effect on a copyrighted work's potential market value (a minuscule but essential portion that drains economic value from the owner isn't fair).[15]

Second, copyright terms eventually expire and the work enters the public domain. At that point, anyone is free to use, adapt, or distribute copies of the work without permission and for free. In the United States, for works created in 1978 or later, copyrights generally last the creator's life plus seventy years. For "an anonymous work, a pseudonymous work, or a work made for hire, the copyright endures for a term of 95 years from the year of its first publication or a term of 120 years from the year of its creation, whichever expires first."[16]

Another important term that applies in the copyright context is "work made for hire." Creative works a person makes in the course of his or her employment, such as writing a report or a newsletter, writing code and algorithms for a mobile or computer application, or drawing sketches or illustrations, belong to the employer. Similarly, an independent contractor, such as a freelance writer, a website designer, or an independent photographer, commissioned or hired to produce certain creative works under contract, creates works made for hire. The employer or one who hired the contractor actually holds the legal rights to the work product. The employer or hiring party directs the work project—tells the creator what to make and particular instructions about its details—and compensates the creative functionary such as through a paycheck, and provides the supplies and other materials as well as workspace and tools to carry out the task. This differs from the writer or artist initiating and controlling a project on her or his own time. The employer may share the work made for hire with the creator but has no obligation to do so.[17]

Patents

Patents in the IP context secure an inventor's exclusive rights to his or her inventions and discoveries. A patent is like a deed to land. Only, this "land" never existed before. And just as with one's artistic, musical, literary, or other creative works, inventive discoveries are the fruit of someone's intellectual labor. The difference is patents apply to "useful arts" such as inventions in mechanics or electronics versus creative works in the "fine arts" such as music, literature, or painting, where copyright applies.

One summary of patents says, "Described as a set of exclusive rights granted by a country or state to an inventor for a limited period of time, patents are subject to requirements of detailed public disclosure of an invention. The invention is a solution to a specific technological problem and may consist of a product or a process of manufacture."[18]

Patent, more generally, is defined as "a grant of some privilege, property, or authority, made by the government or sovereign of a country to one or more individuals."[19] For example, in 1663 British King Charles II issued a land patent that chartered Carolina land holdings to a group of eight Lords Proprietors. These men largely were royalists, most having sided with Charles I, Charles II, or both and in returning the throne to the Stuarts. The grant empowered the Lords Proprietors to govern and make use of the territory, which after a few false starts by the group, Lord Anthony Ashley Cooper's leadership established a permanent settlement in 1670 in what would become South Carolina.[20]

Patents have existed for centuries. In the Old World, patents often amounted to largesse or favor bestowed by a monarch. Thus, inventors who lacked political connections may not have asked the king or the government for a patent on their inventions. Instead, they often tightly controlled knowledge of how to make an invention and tightly controlled those involved in the making. Essentially, artisans operated under trade secrecy. Without disclosure, innovation occurred relatively slowly. Patent property rights of exclusivity, coupled with initial disclosure and eventual entry into the public domain, serve to foster innovation and speed technological progress, as well as wealth creation and faster advancement of the human standard of living.

Three types of patents are provided for in U.S. patent law: **utility patent**, **design patent**, and **plant patent**. A utility patent is the "customary type . . . issued to any novel, non-obvious, and useful machine, article of manufacture, composition of matter or process."[21] A design patent covers "the unique appearance or design of an article of manufacture . . . for both surface ornamentation or the overall configuration of an object."[22] A plant patent is issued to a "person who invents or discovers and asexually reproduces any distinct and new variety of plant, including cultivated sports, mutants, hybrids, and newly found seedlings, other than a tuber propagated plant or a plant found in an uncultivated state."[23]

Moreover, a patent on an invention or discovery is defined as "a grant of right to exclude others from making, using, or selling one's invention and includes right to license others to make, use or sell it."[24] Note the use of the term "right" as opposed to the term "privilege" seen in the general definition of patent. The importance of this distinction will become clear as we continue. Patents also secure the right (there's that important term again) to exclude patent-infringing goods from importation.[25]

The legal right a patent secures is "a right to the exclusive manufacture, use and sale of an invention or a patented article."[26] "Patents—like other IP rights—are personal property, and may be assigned, willed, sold, licensed, or otherwise dealt with by the patentee."[27] Whereas other forms of IP enjoy both criminal and civil protections under the law, patent law leaves civil litigation as the only means of enforcing patent rights.[28] Thus, for copyright, trademark, and trade secrets, law enforcement authorities may prosecute infringement crimes, while patent owners are left to fighting infringers on their own by civil litigation.

U.S. patent terms are now twenty years from the date the application was filed with the Patent and Trademark Office. Also, a one-year provisional patent may precede filing a regular patent application. Provisional patents' advantages include a lower initial filing fee, an earlier filing date for the invention, and use of the term "patent pending" in the invention's early commercial efforts.[29] Another advantage associated with the provisional patent is "it gives you one year to assess the invention's commercial potential before you have to commit to the higher cost of filing a nonprovisional application."[30] Similar to copyright, a patent secures the inventor exclusive rights of limited duration. Once a patent expires, the invention or discovery enters the public domain and is usable by anyone without license.

Harold Lindsell captures the essential importance of patents for human creativity in the useful arts: "Those of inventive mind have their ideas[31] protected by patents. Do not new ways of doing things, or new ideas which have never been thought of before belong to their inventors?"[32] This simple tenet should be so manifestly self-evident as to lie beyond refutation. Unfortunately, many people deny and dispute this truth. More on that later.

Trademark

Trademark is a form of IP that protects legal rights for a word, name, slogan, or symbol a company or organization uses to identify itself as a source of goods or to distinguish its products from similar products. For instance, the "Coca-Cola Company," "Coca-Cola," and "Coke" are all trademarks—one of the company, two of its banner product. Similarly, "service marks" protect a firm's services rather than products. Trademarks include a phrase, logo, name, or a graphic. A trademark provides "a distinctive mark of authenticity."[33] Trademarks secure exclusive rights "to prevent others from using a confusingly similar mark."[34]

Trademarks (™) and service marks (℠) may be registered (which bears this ® symbol) with the U.S. Patent and Trademark Office or the trademark registrars of other nations. There is no registration requirement to use a mark (as long as no one else already uses it), but doing so strengthens an owner's hand in enforcing associated legal rights against anyone selling knockoff products using copycat branding, packaging, or designs.[35] Common law rights may be established in unregistered marks by proving commercial use. One's property rights in his or her trademarks do not expire, as long as the marks continue to be used in commerce and, if registered, are renewed every ten years.

The Lanham Act of 1946 governs U.S. trademark law, and states provide trademark registration within their jurisdictions. The Lanham Act "created the modern U.S. trademark system, providing certification for the first-use date, renewal requirements, defensive registrations, and enforcement provisions."[36]

Trade Secret

Trade secrets are highly valuable, closely guarded business information, such as a formula, process, or chemical compound. *Black's Law Dictionary* says, in part, a trade secret is "a plan or process, tool, mechanism, or compound known only to its owner and those of his employees to whom it is necessary to confide it. A secret formula or process not patented but known only to certain individuals using it in compounding some article of trade having a commercial value."[37]

Choate's definition of trade secret adds important elaboration: "A device, method, or formula that gives one an advantage over the competition and which must therefore be kept secret if it is to be of special value."[38] The economic value of this type of IP lies in its remaining undisclosed and unknown.

Well-known examples of trade secrets include the Kentucky Fried Chicken recipe and the Coca-Cola formula. Those two recipes are worth billions of dollars, support thousands of jobs, and have proven their worth for many, many years—for more than a century in Coca-Cola's case. Other types of confidential, proprietary information closely approximate trade secrets. Know-how, the "knowledge of how to do something,"[39] involving certain operational, research, or other procedures, practices or methods, may be considered trade secrets.[40] Know-how implies not only knowledge, but skill in its application. Proprietary data sets, for example, are often regarded as trade secrets.

Contracts and nondisclosure agreements are the predominant protection of such confidential, proprietary information against misappropriation.[41] Section 337 of the Tariff Act is available for blocking the import of foreign-made products that misappropriate a U.S. trade secret. In addition, the 1996 Economic Espionage Act provides criminal sanction for stealing trade secrets.[42] Also, the 2016 Defend Trade Secrets Act provides a federal civil right of action for trade secret misappropriation.[43] The Supreme Court held in *Ruckelshaus v. Monsanto Company*[44] that a trade secret is property constitutionally protected under the Takings Clause of the Fifth Amendment; therefore, the government owes the owner compensation for its unauthorized use or disclosure.[45]

Trade secrets are perpetual, unless disclosed. They don't expire like a patent or copyright and have no public disclosure requirements. Moreover, competitors are free to try to figure out the original item's secret sauce. "It is legal to use reverse engineering to learn a competitor's trade secret," Choate notes.[46] If reverse engineering leads to discovery of another's trade secret, the competitor who made the discovery bears no liability. The trade secret wasn't stolen, no one with knowledge of the secret violated any contractual obligations to secrecy, and so the owner of trade secret type of IP has little recourse to protect against its unauthorized usage.

● ● ●

This discussion of the primary forms of intellectual property—copyright, patent, trademark, and trade secret—leads into our focus on patents for further considering creativity and ownership. We begin this in-depth biblical perspective on invention and patents with this reminder: Each type of IP has its distinctives, as do other forms of private property and property rights. The common denominator of the various kinds of IP examined above is that each secures private property rights in newly created property. There would be no such property to use, make, own, and put to use had some person not created it. These species of title to property merely secure legal rights in one's creations.

Part III:
THE PATENT ECOSYSTEM

Chapter 14

The American Patent System and the Iconic Inventor

> *"The copyright of authors has been solemnly adjudged, in Great Britain, to be a right of common law. The right to useful inventions seems with equal reason to belong to the inventors. The public good fully coincides in both cases with the claims of individuals."*
> — James Madison, Federalist 43

> *"Who does not consider with sincere esteem, and bestows their warmest approbation, on those who have advanced the useful Arts and Sciences, or contributed to the support of true virtue and religion?"*
> — Bishop Robert Smith, 1779 sermon[1]

Once shots were fired between American minutemen and British troops in 1775 at Lexington and Concord, mounting tensions over the Mother Country's maltreatment of its thirteen colonies along North America's Atlantic coast escalated into the Revolutionary War. The American Revolution, pitting American blue coats against British red coats, lasted beyond the 1781 American victory at Yorktown, Virginia, lingered as British troops remained on American soil until 1782, and finally ended with the signing of the Treaty of Paris in 1783.[2]

The eight-year struggle for independence from the world's dominant empire, contested against its powerful army and navy, exposed serious weaknesses that the newly independent American states would need to address—and fast. How could a mostly agricultural society lacking industrial strength supercharge its economy and industrial progress?

The American Founding and Intellectual Property

The Continental Congress was comprised of notable Americans, including Benjamin Franklin and future U.S. President Thomas Jefferson. Among them were future President John Adams, his cousin Samuel Adams, and John Hancock from Massachusetts; Connecticut's Roger Sherman, signer of both the Declaration and the Constitution; the Rev. John Witherspoon of New Jersey; Elbridge Gerry, a Declaration delegate for Rhode Island and a Constitutional convener for Massachusetts; and Pennsylvanian Benjamin Rush.[3]

To compose a Declaration of Independence, the Congress selected thirty-three-year-old Jefferson to serve on the committee. Jefferson chaired the committee and was tasked with writing the Declaration's first draft. Early in July 1776, Jefferson's stirring document was modestly amended, adopted, and signed by all fifty-six delegates.

At the time of the Constitutional Convention, former Virginia Governor Thomas Jefferson was American minister in France. When Jefferson returned to the United States in October 1789, President George Washington offered Jefferson the job of secretary of state; Jefferson accepted.[4]

In 1787, the Articles of Confederation having proven lacking as a model of government, the thirteen former British colonies formed a fledgling nation under a written constitution. They won their independence, despite chronic shortages of such essentials as guns and gunpowder, shoes and blankets, other necessary goods, and war materiel. The patriots commonly fought a well-supplied enemy with around nine cartridges each for the Americans' muskets. "At the battle of Bunker Hill, the Americans quickly ran out of ammunition, finishing the fight by clubbing the English troops with the butt ends of their muskets."[5] As Choate points out, General George Washington's troops suffered acutely from the bitter cold at Valley Forge, Pennsylvania, that 1777 to 1778 winter. Thousands contracted illness and frostbite, while many died.

In the more than a century and a half of colonial American society, a somewhat virtuous economic circle developed, the colonies trading with their colonial motherland, supplying raw materials Great Britain shipped back as finished goods. The industrial powerhouse provided a market for America's produce, while America provided Britain a market for the goods the Mother Country made. A sort of win-win for a while later became problematic—especially when war broke out.

This problem was recognized early on. In 1775, Benjamin Rush observed, "A people who are dependent on foreigners for food or clothes must always be subject to them."[6] Combat brought crisis for the would-be free colonies. "From the beginning of the war, Washington's army lacked guns, gunpowder, rope, sails, shoes, and clothes, among many other military necessities, largely because Great Britain had long prohibited most manufacturing in its American colonies," Choate writes. "Instead, the mother country restricted colonial production to timber, furs, minerals, and agricultural goods."[7] The economic and military crisis the war brought on caused Americans to recognize the need for changing the model to promote economic and technological progress.[8]

* * *

When the Constitutional Convention convened in Philadelphia in 1787, some of the delegates aimed to foster creativity that could translate into practical benefits for the newly independent nation. Promoting invention was to become a priority of the newly constituted government, having learned the lessons of the Revolutionary War's hardships getting materiel for waging a war. The Founders made invention a key to sparking America's economic diversification and rapid growth. The Constitution they framed empowered Congress to secure intellectual property rights—notably, the only place in the original Constitution where "right" appears.

Of the fifty-five delegates to the Constitutional Convention, only eight—the best-known being Benjamin Franklin of Pennsylvania—had also served in the second Continental Congress which produced the Declaration of Independence eleven years before. Franklin served on the committee tasked with drafting the Declaration, which he voted for and signed. He spent much time abroad, before the Revolution in England as Pennsylvania's representative and during the war as American envoy to France. Upon war's end, Franklin negotiated and signed the Treaty of Paris, in which England agreed to peace with an independent America. In

1785, Thomas Jefferson relieved Franklin. Franklin returned to Pennsylvania, which in 1787 made him a Constitutional Convention delegate.[9]

* * *

In chapter 12, we examined the Lockean principles of property rights, drawn from biblical truths and Common Grace, and Jefferson's weaving them into the Declaration as "certain unalienable Rights" listed as "Life, Liberty, and the Pursuit of Happiness." We further saw inalienable rights, such as private property and safety, provided for in the U.S. Constitution. We now consider intellectual property in the Constitution.

If the second Continental Congress counted leading lights as members, the Constitutional Convention was equally flush with statesmen. It included George Washington, James Madison, Alexander Hamilton, Franklin, and George Mason among others. But not Thomas Jefferson, who was in France.

Franklin[10] and Jefferson were exceptional men with exceptional minds who lived exceptional lives. Both were science enthusiasts and inventors. Franklin invented, for example, the lightning rod, bifocal eyeglass lenses, and a more efficient stove that used less fuel while producing more heat. Jefferson invented, for instance, a more efficient, easily reproduced moldboard plow, which is "the part of the plow that lifts up and turns over the sod cut by the iron share and coulter,"[11] a wheel cipher design for composing and reading coded messages,[12] and a zigzag tin roof.[13] Neither patented his inventions. This sounds especially strange for Jefferson, the first U.S. patent commissioner, then a duty of the secretary of state. Franklin publicly disclosed his inventions, including his stove invention.

Jefferson, a follower of Locke on property rights, objected to any type of monopoly, patents among them.[14] In 1790, Jefferson appears to have warmed up to patent rights: "An act of Congress authorising the issuing patents for new discoveries has given a spring to invention beyond my conception. Being an instrument in granting the patents, I am acquainted with their discoveries. Many of them indeed are trifling, but there are some of great consequence which have been proved by practice, and others which if they stand the same proof will produce great effect."[15]

However, in an 1813 letter responding to whether a natural right in inventions exists, Jefferson asserts inventions cannot be confined or exclusively appropriated: "Inventions then cannot, in nature, be a subject of property."[16] Jefferson considered it "singular to admit a natural and even an hereditary right to inventors."[17]

Erler explains Jefferson's position—believing it aligns with Locke's—a view that denies natural right to one's creations, but "occupation and improvement creates the right to property." Erler continues, "Whatever property a man creates is for his use and convenience, and this means that he may extinguish the property by use or exchange, thus alienating the property he has created without alienating the natural right to property itself."[18] In this view, because the creation may be disposed of or worn out, ceasing to exist, though private property, it is less than by natural right.

Jefferson's and Franklin's attitude toward the ownership rights of an individual's creative output seem not widely shared by their fellow statesmen or by those who invest themselves in inventive and creative endeavors.

* * *

At the Constitutional Convention, James Madison, the "Father of the Constitution" and future U.S. president, engaged in the process and the debate. Madison's commitment to property rights extended to intellectual property (IP) rights. The Virginian Madison and Charles Pinckney III, a delegate from South Carolina, proposed the IP clause found in Article I, Section 8, Clause 8.[19] This section of the Constitution authorizes specific powers of the U.S. Congress. The provision was brought up in convention on September 5, 1787, among a handful of others. The Madison-Pinckney inventors' and authors' provision was amended only by inserting one word in the text. No debate occurred on this measure. The convention adopted the IP clause without opposition.[20]

Placing intellectual property in the Constitution was "unprecedented," legal scholar and IP expert Adam Mossoff observes. "No country's founding document had done this before" (internal footnotes omitted).[21] Those at the Constitutional Convention and delegates to the state conventions considering the Constitution's ratification must have recognized this as a novel move, placing intellectual property on the same plane as other forms of property as a matter of foundational law. And yet doing so caused no extended or controversial debate of the clause—in fact, no debate at all. It likely helped that several American colonies and, after independence, states adopted patent and copyright laws similar to Britain's Statute of Monopolies and Statute of Anne, respectively.[22]

Indeed, the novelty of the American IP model was intentional. "Just as the U.S. 'implemented innovative structural and substantive changes in its new political and legal institution' after the American Revolution, the patent system authorized in the Constitution and implemented with the first Patent Act of 1790 'represented the same fundamental break from the English patent system as other U.S. political and legal institutions.'"[23]

The Founders understood "patents and copyrights encouraged inventors and authors to produce more new and useful creations. These innovations could help the U.S. progress. And as the details of these creations became public, the general knowledge of the nation would be expanded." This would result in making the new United States of America "grow richer and stronger faster."[24] Incentivized, rights-secured creativity and ownership of new technologies, inventions, etc. would help overcome the kinds of shortages and travails American patriots had faced.

From this genesis, the American model for a patent system began to take form and take root. New York's City Hall was refurbished into Federal Hall to seat the newly constituted United States government. In addition to space for Congress to meet and a library, Federal Hall contained a room to house models of inventions.[25] The "machinery room" represents the forward thinking of the Founders and their faith in the fruits of the labor patents and copyrights would stimulate through secure, exclusive ownership.

Further evidence is the fact President Washington, in the first State of the Union Address January 8, 1790, urged the First Congress to enact a patent law. Congress did so three months later. The third U.S. law enacted was the Patent Act of 1790, followed just weeks later by the Copyright Act of 1790.[26]

The Constitution's Foundational IP Features

The Founders' noncontroversial IP clause, which wrote patents and copyrights into the text of the U.S. Constitution, not only laid the foundation for the first American patent and copyright laws, it led to important statutory elements distinguishing U.S. patents and inventions as private property. Article I, Section 8, Clause 8 empowers Congress "to Promote the progress of Science and useful Arts, by securing for limited Times to Authors and Inventors the exclusive Rights to their respective Writings and Discoveries."

In Federalist 43, one of the essays Alexander Hamilton, John Jay, and James Madison penned arguing for adoption of the Constitution to replace the Articles of Confederation, Madison devotes two paragraphs to justify the IP power, the second saying:

> The utility of this power will scarcely be questioned. The copyright of authors has been solemnly adjudged, in Great Britain, to be a right of common law. The right to useful inventions seems with equal reason to belong to the inventors. The public good fully coincides in both cases with the claims of individuals. The States cannot separately make effectual provisions for either of the cases, and most of them have anticipated the decision of this point, by laws passed at the instance of Congress.[27]

Madison notes the unquestioned beneficial nature of the power enumerated. The provision adopts the copyright-as-a-property-right precedent and extends the same legal protection to the fruits of inventors' labors. He writes that IP exclusivity serves both public and private interests. State laws offer only so much protection of these property rights, necessitating a federal IP law.

* * *

Four features of the IP clause warrant highlighting. Through them, the Founders fashioned the original "open-source" innovation model. They advanced a public purpose. They leveraged private property rights. They democratized IP. They incentivized sharing one's discoveries.

First, the clause contains a ***purpose statement*** as its grant of power: "to promote the progress of science and useful arts." This aims to channel Americans' application of their intellectual and creative faculties in their respective fields to expand the body of knowledge and to create useful items.

By "useful arts," the Founders sought new discoveries and inventions of practical use. These involve skilled workmanship or human intervention—making new or improved implements profitable, beneficial, efficacious. This practically applies new knowledge. Today, we think of it as applied research and development, or R&D, stemming from basic research findings.

Second, this provision employs ***private property rights*** to achieve both private and public purposes: "securing for limited times . . . the exclusive rights to their . . . writings and discoveries." As Madison says in Federalist 43, "The right to useful inventions . . . belong to the inventors. The public good fully coincides . . . with the claims of individuals." The phrase recognizes private property interests. Congress secures authors' and inventors' property rights to the fruits of their intellectual labor. These are individual rights. Mossoff explains, "Consistent with the understanding at the time that the role of the government is to define and secure individual rights, Congress is authorized here to secure a right of individual inventors and authors" (internal footnotes omitted).[28]

Moreover, an invention is rightly understood as private property beginning from its discovery. Chief Justice John Marshall calls an unpatented invention "inchoate and indefeasible property" that's "vested by the discovery."[29] The patent merely secures property rights that preexist the patenting process; ownership rights attach upon invention, not patenting. Marshall explains the "constitution and law, taken together, [secure] to the inventor, from the moment of invention, an inchoate property therein, which is completed by suing out [i.e., applying for and obtaining] a patent."

Patents secure rights to property that didn't exist before. The U.S. government's role is to ascertain (1) the true inventor of the newly created property, (2) that the creative work is new and useful, and (3) the new property's boundary lines. Also, the exclusive property rights, once patented, are of limited duration. Initial exclusivity allows inventors and writers to reap the first-fruits of their labor.

Third, the American system ***democratizes IP rights***.[30] The clause implies newly created property belongs to its originator, just as Chief Justice Marshall affirms. It uses the personal pronoun "their" as in "their discoveries." The Patent Act of 1790 refers to "the first and true inventor."

Our democratized patent system differs from Old World practices. Because sovereigns awarded patents to elites for a price, common craftsmen and artisans often protected their inventions as trade secrets. But in America, the first and true inventor received the patent on his or her invention, regardless of his or her station in life.

Phoebe Cade Miles, cofounder of the Cade Museum for Creativity and Invention and daughter of the lead Gatorade inventor, elaborates on the democratization of IP rights in the United States. The U.S. Patent Office "was gender blind, and it was colorblind. Women, before they had the right to vote, had the right to invent, and husbands were not allowed to sign their patent application. Only the true inventor was allowed to do that."[31]

Fourth, the open-source aspect of the Founders' approach to innovation provides **public benefit by maximizing private interest**. Inventions' and creative works' IP provides an individual exclusive rights. But those exclusive IP rights last only for a limited time. Inventive and creative works eventually become public. When a patent or copyright term expires, claims of private rights no longer hold. The work enters the public domain. Others may then freely make or use the invention or creation.

Further, part of the tradeoff for initial exclusivity is public disclosure once a patent is granted. The 1790 Patent Act required inventors to provide detailed specifications and a model or drafts that "enable a workman or other person skilled in the art or manufacture . . . to make, construct, or use the [invention], . . . [so] the public may have the full benefit . . . after the expiration of the patent term."[32] Thus, exclusive rights don't freeze progress until the invention's entry into the public domain. The new knowledge shared immediately enables progress to advance the state of the art while the patent remains in force.

Exclusive patent rights, coupled with immediate public disclosure and the invention's eventual entry into the public domain, is called the Patent Bargain: The inventor gets exclusivity for a time. The public gets immediate learnings from patented inventions and the freedom to invent around the invention's metes and bounds. And, eventually, the public gets free access to the invention.

In other words, the American IP model combines in a novel, brilliant way private property rights, public purpose, democratization of patent grants, and shared learnings. Meanwhile the inventor or creator enjoys the full fruits of his or her labor while society enjoys the benefit of the product, implement, device, or other creative output—a new feature film, a better medicine, a technological advance in telecommunications or aeronautics, or a new song or book.

The Great Awakening and the Founding Generation

Secure, exclusive IP rights, such as those embodied in the U.S. Constitution and the Patent Acts of 1790, 1793, and 1836, align with biblical and Common Grace principles of human creativity and ownership.[33] Was this a coincidence, or may religious influences have helped shape the Founders' perspective on invention and patents as property?

The Founders were reared in a Judeo-Christian culture. They grew up where Christianity and churches (of different varieties, mostly Protestant) were generally prevalent in society. There were also Catholics and Jews,[34] though most Americans during colonial times shared a "common Christian heritage." Individuals ranged from being "nominally, habitually or devoutly" religious. "Virtually all" Christians believed God to be "Creator, Provider and Disposer," that life is split between "Time and Eternity," Jesus Christ is God entering history as a fully divine human being to provide saving grace, and the Bible is God's message to humans.[35]

That's not to say America was consistently a "shining city on a hill," as far as personal piety goes. Not long after the Pilgrims established the Plymouth colony in the 1620s, religious belief and church attendance dipped to 5 percent of the population, while "extremely low moral conditions prevailed."[36] By late October 1739, when the Rev. George Whitefield arrived in America to preach, "the religious fervor which had characterized many of the first settlers of the new world had long since died away."[37] In 1706, the Rev. Cotton Mather decried "a general and an horrible decay of Christianity. . . . The modern Christianity is too generally but a very spectre, scarce a shadow of the ancient."[38] The fields were ripe for harvest.

. . .

The Bible was widely available, commonly read in both religious and secular settings in public and private, and was generally revered in the colonies.

"The Bible was the most prominent literary text in eighteenth-century America," American University scholar Daniel L. Dreisbach writes. "The culture was religious, and the American people were biblically literate."[39] The Bible was central beyond the religious to the Founding generation's education, "intellectual pursuits," arts, public matters, and law.[40]

In addition to Scripture, other religious writings were published in the eighteenth century: "sermons, tracts, practical guidebooks, and Biblical commentaries."[41] The most popular colonial American authors included the Rev. Jonathan Edwards, Benjamin Franklin, and the Rev. Cotton Mather, the seventeenth- and eighteenth-century New England Puritan minister.[42] Also, Franklin published popular English preacher George Whitefield's sermons in America.[43]

Some Founders considered the Bible to be God's Word. Dreisbach cites several, such as John Jay, signatory to the Constitution, *Federalist Papers* cowriter, and the first U.S. Supreme Court chief justice; Robert Treat Paine, Declaration of Independence signer; Roger Sherman, who signed the Declaration, the Articles of Confederation, and the Constitution; and John Witherspoon, a Presbyterian minister, college president, and signer of the Declaration.[44]

Scripture permeated the Founders and their political efforts. Indeed, "notwithstanding Enlightenment influences on intellectual elites, the founders cited the Bible more frequently in their political discourse than any other work."[45] Biblical citations outpaced legal and political philosophers Blackstone, Locke, and Montesquieu with the Founders.[46] John Adams recognized the biblical influence and its implications for public morality in a republic. In an 1807 letter to Benjamin Rush, Adams writes, "The Bible contains the most profound philosophy, the most perfect morality, and the most refined policy, that ever was conceived upon earth. It is the most republican book in the world, and therefore I will still revere it."[47]

• • •

During the early to middle decades of the eighteenth century, the Great Awakening shaped the culture and society into which the Founders were born or bred. This Christian revival movement swept the American colonies from the 1720s through the 1740s—formative years for the Founding generation. "Religion was important to . . . Calvinists and evangelicals, formerly 'dissenters' to the established church. Many, like [SC militia General Andrew Pickens, b. 1739], had been born or come of age while the fires of the Great Awakening were still smoldering in the American colonies . . ."[48] Ministers Jonathan Edwards in New England, William and Gilbert Tennent in New Jersey and the Middle Colonies, popular itinerant preacher George Whitefield throughout the colonies, and other evangelists took the Gospel of Jesus's saving grace up and down the Atlantic seaboard.[49]

This widespread, sustained evangelical revival turned many, many Americans, from a broad range of sects, especially Protestant denominations, throughout the colonies to faith in Christ and many to a life of piety in their newfound faith in the Savior. The ministers were "dissenters," those in denominations, such as Baptist or Presbyterian, outside the "established," or official, state church. Many held to Reformed doctrine.[50] They faithfully preached the Bible, both Old and New Testaments, and the Gospel of Christ, while the Holy Spirit awakened the dead souls of their hearers.

The Great Awakening's effect was profound. The phenomenon marked a widely shared experience, providing certain commonality among the disparate people of the thirteen disparate colonies. The fact that Whitefield's clear, booming voice was heard by Americans from Georgia into New England in churches, homes, streets, and backwoods gave people, from all walks of life and religious backgrounds, including Native Americans and slaves, the common touchstone of Whitefield's commanding delivery speaking the Word of God and the Gospel of Christ. Dallimore says, "Far more than half of the total population of the Colonies had heard [Whitefield] preach."[51]

A large share of Whitefield's hearers, through Christ, gained a heart of flesh replacing their heart of stone. The fruits of his and other evangelists of the Great Awakening proved genuine in many Americans. Regarding those fruits of Whitefield's 1739 to 1740 trip to America is a partial summary:

> [After Whitefield's preaching visit,] Bibles were in greater demand than ever before and a large market for religious books had now

come into being. . . . Hosts of people who had come to the full assurance of salvation now thronged the evangelical churches and gathered in groups in homes and barns to study the Scriptures and pray. The figure was given of thirty Societies that had been established in Boston and there were also several—Societies for children, Societies for young people, Societies for women, Societies for black people and Societies for black and white people together—in Philadelphia, and much the same was true of New York. . . . All of these were related to the local churches, and . . . Whitefield not only made no attempt to organize them as his own work, but that he was strongly averse to any suggestion of a new denomination or a personal party of any kind.[52]

This revival "cut across the bounds of colonies and denominations to provide a common ground in evangelical religion to many inhabitants throughout the colonies."[53] The Great Awakening contributed to an already developing and spreading religious tolerance in the colonies. Additionally, America's "first truly *mass* movement," while religious, "was to be felt in the politics of the era" (emphasis in original).[54] The Awakening of souls "encouraged the perception of a human as an individual—an individual of choice and dignity—. . . [who seeks] freedom in government, as well as freedom in religion."[55] It "prepared the way for religious liberty."[56] Thus, the Bible and Christianity, particularly at the time of and following the First Great Awakening of the eighteenth century, played a significant part in shaping the Founders' worldview, including their understanding of creativity and ownership.

* * *

What might the Bible teach the Founders that could apply to patents and invention, creativity, and ownership? They would have been familiar with many of the Bible passages we have looked at in our study here.

First, Scripture teaches about God's creativity. The Bible instructs how human beings are God's image-bearers, possessing faculties for creativity. There's also the Creation Mandate, commanding humans to apply their abilities and flourish.

The Bible shows God is the divine Creator. He made the universe and everything in it. The Founders learned, "In the beginning, God created the heavens and the earth," and he's making the new heaven and new earth. From plants to planets, atoms to electricity, all things are God's creations.

Second, Scripture calls God the owner of his creative works. He lays claim to everything he made. The Founders learned the earth and everything in it belong to the Lord. They saw how the Ten Commandments are about property rights. The first four commandments prohibit theft of God's property, the last six command respect for other people's private property. God's ownership of what he creates flows naturally from his having created the things owned.

Inescapably, Scripture closely connects creating something and owning it. The Bible repeatedly affirms God's inherent right to own what he's made. And the Founders would have learned the analogy of the potter and the clay. A creator has the right to determine what to create, how it's made, and its use. The Founders would have recognized ownership as inherent.

Third, the Bible teaches that humans are made in God's image; we are rational and spiritual. We're God's highest creature, male and female. Thus, the Lord endowed humans with certain communicable attributes, including creativity and ownership.

Such characteristics are part of his Common Grace. The Lord sends rain and sunshine on the righteous and the unrighteous alike. The laws of nature—such as of science and mathematics—are, as Galileo put it, "written by the hand of God."[57] They're generally knowable, discoverable by humans. This falls within Common Grace as part of God's general provision.

Fourth, Scripture teaches the Creation Mandate. It preceded the Fall from grace into sin and applies to all humankind. God's mandate tells us to be fruitful—that is, to steward plants, animals, and nature. God commanded humans to work. We're to fill and subdue the earth. The Founders would have understood God instructs humans to form families, to put to use the faculties and natural resources he has provided, and to tend his creation. This, too is a part of general revelation, applying to all individuals.

In this model, like us, the Founders would see that God provides for his creation through Common Grace and the Creation Mandate. God gives talents and abilities to saints and sinners alike. We're each to use the talents, gifts, abilities, skills, and interests given us. We own what we create, and what we make may benefit society. Human work brings about human flourishing. This resembles God's calling his completed creation "very good."

For the Founders, the Bible was quite familiar. Their society was much more fully cognizant of Scripture; even the illiterate were far more biblically literate than many educated people of the secular twenty-first-century West. Biblical teachings about creativity and ownership would have resonated with the Founders, many regarding them as self-evident truths. Many Founders would apply such truths about creativity and ownership to intellectual pursuits.

"What Hath God Wrought!"

America's model for a patent system, informed by biblical principles, prevailed for about 200 years. This patent system led the United States to an unprecedented innovative leap in the nineteenth century and beyond. In one century, America went from upstart nation with an agrarian economy to the world's industrial and innovation leader. American inventors, empowered by secure patents, created new industries and revolutionized others through their inventions—for instance, electrification, wired and wireless communications, railroads, steamboats, agriculture, motion pictures, manufacturing.

The nineteenth century saw American inventors from all walks of life create all kinds of inventions and advance the useful arts. The "assiduous merchant, the laborious husband, the active mechanic, and the industrious manufacturer" whom Alexander Hamilton in Federalist 12 aims to prosper through the new Constitution applied their faculties in inventive ways.[58] Tocqueville, the Frenchman who recorded in *Democracy in America* his impressions of the young nation in the New World, observes, "Their strictly Puritan origin, their exclusively commercial habits, the country they inhabit, which appears to divert their minds from the study of science, literature, and the arts, . . . must have focussed [sic] the American mind . . . upon purely practical concerns."[59]

The people of the American colonies and subsequent unified states had both practical and, for many, religious priorities.[60] Practically, they had to build a home and a community. They needed food, shelter, protection against wild animals, Indians, and other dangers. They needed to establish commerce and develop trade and an economy. Especially for those on the edges of settled territory, these concerns stood foremost.

Certain aspects lent themselves to common Americans' pursuit of happiness in practical ways. Landes, in *The Wealth and Poverty of Nations*, describes these:

> America's society of smallholders and relatively well-paid workers was a seedbed of democracy and enterprise. Equality bred self-esteem, ambition, a readiness to enter and compete in the marketplace, a spirit of individuals and contentiousness. At the same time, smallholdings encouraged technical self-sufficiency and the handyman, fix-it mentality. Every farm had its workshop and anvil, its gadgets and cunning improvements. Ingenuity brought not only comfort and income but also status and prestige. Good workers were the envy of their neighbors, the heroes of the community. Meanwhile high wages enhanced the incentive to substitute capital for labor, machines for men.[61]

Certainly, living on or near a farm, a blacksmith's shop, or a foundry, as well as ample need for family members to pitch in and help with chores and in urgent circumstances, provided Americans, from cities and towns to the backwoods, with practical experience. Learning skills and applying oneself to finding practical, workable solutions to everyday problems would instill confidence and enhance ability.

The arrival of the American patent system expanded the incentives for ingenuity beyond solving an immediate problem: "There is not a working boy of average ability in the New England states, at least, who has not an idea of some mechanical invention or improvement in manufactures, by which, in good time, he hopes to better his position, or rise to fortune and social distinction."[62]

• • •

The list of nineteenth century inventors who became household names is impressive, as were their inventions. For instance, Eli Whitney's cotton gin, patented in 1794, was disruptive technology, making cotton king in the South and textile manufacturing prosperous up North. Samuel Morse's telegraph, patented in 1840, used "Electro Magnetism" to send signals containing information by wire. Morse, who created the alphabetical Morse code of "dots" and "dashes," transmitted the first telegraph message on his invention in 1844, from Washington, District of Columbia, to Baltimore, Maryland. Morse tapped the encoded phrase from Numbers 23:23, "What hath God wrought," which his associate in Baltimore transmitted back.[63] Alexander Graham Bell's telephone, patented in 1876, enabled transmission of sound over electric wire.[64]

Lesser-known nineteenth-century inventors include Helen Blanchard, who invented several improvements to the sewing machine. Blanchard's first patented invention in 1873 varied the depth of needle stitches to sew a zig-zag pattern. This holder of twenty-eight patents is in the National Inventors Hall of Fame (NIHOF). Miriam Benjamin, the second black woman granted a patent, invented a gong and signal light device for hotel chairs. Guests press a button on the chair, which rings a gong and turns on the light. Hotel attendants respond to the sound and by the light know which guest is summoning service. Benjamin's chair device was patented in 1888.[65]

Margaret Knight, then known as the "woman Edison," in 1867 invented and later patented the flat-bottomed paper sack commonly found in stores for customers to carry their groceries or other purchased items. Knight owned twenty-seven patents and was a prolific inventor of other, unpatented inventions. Her inventions included a machine to cut shoe soles, a numbering machine, and several inventions related to rotary engines. Knight founded and built the Eastern Paper Bag Company around her patented invention. She went to court to protect her bag-making invention and her business, winning an injunction against a patent-infringing rival bag company.[66]

The U.S. legal system provided inventors, creators, and IP owners certainty, reliability, and enforceability of their patents and copyrights—on par with other forms of property. Litigation over patents and other IP became routine, "a rite of passage common to the more successful inventors of that era."[67] Private property rights, as the Founders intended, prevailed. "By the end of the 19th century, U.S. institutions and legal rules securing patents as property rights in technological innovations had become the 'gold standard' for the rest of the world," Mossoff writes.[68]

• • •

The pace of technological progress and the tangible benefits they brought captured the interest and imagination of Americans in that era. "There is no clinging to the old ways; the moment an American hears the word 'invention' he pricks up his ears," a visiting German said.[69] Famous American author, humorist, and inventor Mark Twain in his 1889 book, *A Connecticut Yankee in King Arthur's Court*, has his protagonist say, ". . . the very first official thing I did in my administration—and it was on the very first day of it too—was to start a patent office; for I knew that a country without a patent office and good patent laws was just a crab and couldn't travel any way but sideways and backwards."[70]

In 1842, a building specifically erected to house the U.S. Patent Office became the office's home. The Patent Office, located seven blocks east of the White House, remained there until 1932.[71] That span of years makes this structure, now part of the Smithsonian Institution's museums, the place where iconic American inventors such as Edison, Westinghouse, the Wright Brothers, and Bell secured their ownership rights in their iconic inventions.

Perhaps most indicative of America's pride in its inventive leadership is collective portraiture titled *Men of Progress*. The 1862 painting of nineteen American inventors depicts most of them seated at a table and several standing nearby. Samuel Morse sits in the middle of the fictionalized gathering. His telegraph, which changed communications significantly, sits on the table by its inventor. It's clearly the group's focus. A portrait of an earlier famous American inventor, Benjamin Franklin, hangs on the wall behind them.[72]

The U.S. Patent Office was housed at this building on F Street, NW, in Washington from 1842 until 1932. The structure is now part of the Smithsonian Institution, in which the American Art Museum and National Portrait Gallery reside. (Public domain. Smithsonian Institution.)

Men of Progress, *painted in 1862 by Christian Schussele, features portraits of nineteen prominent American inventors of the "golden age of American inventing." It captures in a fictional scene the country's favorable sentiment toward invention and patents. Samuel F. B. Morse, inventor of the electric telegraph and a portrait painter, sits in the middle with his invention on the table. (Public domain. Smithsonian Institution.)*

By 1899, the magazine *Punch* could print a comedic conversation about the coming twentieth century. Part of it goes: "Isn't there a clerk who can examine patents?" "Quite unnecessary, Sir. Everything that can be invented has been invented."[73]

Cumulatively, from 1790 through 1835, the United States granted 9,433 patents. In 1869, the 100,000th U.S. utility patent was issued. By 1899, just thirty years later, American invention had added over a half-million more patents.[74] In the twentieth century, we marked 1,000,000 patents in 1911 and 5,000,000 in 1991. The ten millionth U.S. patent was issued in 2018. In 2021, we passed the eleven-million mark.[75]

* * *

In short, to answer Samuel Morse's question, "What hath God wrought," in regard to American ingenuity powered by individual creativity and ownership in fostering progress of science and the useful arts, the answer is simple. God wrought much creative output, practical benefit, prosperity, and human flourishing, thanks to the American Founders' faithful application of biblical principles of creativity and ownership, and their clear vision.

The Founders' patent system resulted in widespread human flourishing. For all their inventing and patenting, the McCormicks, Morses, Edisons, and Bells were only getting U.S. invention started. There was so much more invention to come.

Chapter 15

The Modern U.S. Patent System

"Progress in knowledge and innovations was the primary factor behind the growth of America's economic productivity. . . . The Founding Fathers created a system of innovation that has worked better than any other in the history of the world."
— Pat Choate[1]

"It must be remembered that today, when everyone knows the Wright invention, and the world has assigned certain words to describe it, these words now produce a mental picture which they did not and could not produce when men knew nothing of this method of [airplane flight] control."
— Wilbur Wright[2]

The U.S. patent system, as reformulated by the Patent Act of 1836, catalyzed American invention and patenting. It turned out *Punch* magazine's tongue-in-cheek prediction was wrong, not "everything that can be invented ha[d] been invented" in the nineteenth century.[3] In fact, the pace of progress in the useful arts got faster as the twentieth century, also called the American Century, went along. About 700,000 U.S. patents were granted by the year 1900. By the turn of the next century, ten times as many U.S. patents had been issued, about 7,000,000 by Y2K, the year 2000.[4]

The modern U.S. patent system has proven quite effective, securing inventors' private property rights for a time, while advancing practical knowledge and useful implements of economic value. U.S. patenting 1.0, the 1790 Patent Act, proven inefficient. Having a patent examination panel comprised of cabinet officers who reviewed patent submissions one day a month was destined to hold up progress. This procedure couldn't keep up with demand.

U.S. patenting 2.0, the 1793 Patent Act, proved too permissive. It had inventors register their inventions with the State Department, without meaningful examination. Inevitable patent disputes landed in court, which would sort things out after the fact: Who is the first and true inventor? What are the invention's metes and bounds?[5]

U.S. patenting 3.0, the 1836 Patent Act, struck a balance and bolstered patent rights. It put in place a new and improved patent examination process, with knowledgeable examiners (instead of cabinet members) assessing claimed inventions to confirm their newness and utility and awarding a patent to the first person who produced the novel invention.[6] The 1836 law established a dedicated Patent Office with an extensive patent library. Patent applicants gained the right to appeal denial of a patent to impartial chief examiners, including *de novo*, or fresh, review in federal court. Patents carried terms of 14 years of exclusivity from the date of patent issuance, with the possibility of a 7-year term extension to ensure an inventor "reasonable remuneration for the time, ingenuity and expense bestowed" in developing the invention and bringing it to fruition and market. Khan notes how U.S. courts and the U.S. Patent Office followed rules-based procedures and transparency in decision-making in patent matters; Choate cites characteristics such as defined legal rights, access to the judiciary, and "a body of precedents" to guide judicial decisions.[7]

The core of the Patent Act of 1836 has largely remained the backbone of the American patent system. There were hiccups in the twentieth century before a renaissance in its final years. In 1952, Congress "clarified and simplified . . . and removed redundancies" in U.S. patent law.[8] However, Congress preserved the basic framework that served iconic inventors—many of whom came from humble circumstances—as well as more corporate invention models. Both flourished under that set of patent laws. Historical data from 1790 to 1920 reveal "the distribution of both inventions and inventors in the United States was more 'democratic' than in Europe."[9] America's patent system, as adjusted in 1836, achieved the balance of property rights and public benefit the Founders intended.

Clear title for the original inventor to her or his invention and exclusivity for a set time period, combined with enforceable rights, due process, the rule of law, and a free enterprise economy, gave patent owners certainty in their intellectual property and access to free markets. To be sure, America's patent system continued to develop throughout the nineteenth and twentieth centuries. Evolutions in the law, both by legislative amendment and judicial application of laws and earlier cases' principles to specific cases (common law), the U.S. economy's maturation, and rapid technological advancement continued, bringing about modifications to the U.S. patent system. However, the fundamental facets of patenting, economic activity involving IP as intangible assets, and the premises of private property rights in one's creative output generally persisted in the United States.

The key 1836 elements, distinguished by awarding patents on the basis of objective merit and respect for patent rights as private property rights—the embodiment of the biblical principles of creativity and ownership—stayed intact for more than a century.

The American Patent System in Its Golden Age

The majority of great inventors [of America's "golden age" of patents and invention] chose to patent their key inventions. The American patent system facilitated the entry of relatively disadvantaged individuals into the field of technology, enabled them to specialize in invention, mobilize resources to fund patenting and commercialize their discoveries and enhanced the diffusion of information and inventions. Patent rights comprised secure assets that were extensively traded and gave inventors with only modest resources the opportunity to appropriate private returns as well as to make valuable contributions to society (internal footnotes omitted).[10]

Thus do patent scholars Zorina Khan and Kenneth Sokoloff characterize the American patent system, U.S. patenting 3.0. They describe the effects of the unique American patent system.

An overview of the modern U.S. patent system, that which existed from the mid-nineteenth century into the twentieth century—what we'll call the "golden age of the American patent system"—will lay the groundwork for subsequent chapters. We briefly examine concepts such as patent eligibility, examination, validity, and assertion. And we highlight important developments of the 1980s and 1990s.

* * *

Before beginning the process of patenting an invention, one must keep his invention to himself, only sharing it confidentially with his coinventors. If information about an invention becomes publicly known before filing a patent, that makes the invention ***prior art***. Previous states of a given art, patent-protected or not, are not patentable.

A prudent step is to perform an initial patent search to ascertain whether someone has already invented and patented the invention intended for filing. As an author, I did a similar search of books already published to see if this book had been written. (Thankfully, it hadn't.) Yonover and Crowe suggest doing an online search, such as of the U.S. Patent and Trademark Office's (PTO) website, particularly its advanced search function or with other search engines.[11]

If this preliminary look doesn't uncover "your" invention as already patented, the next step is to write and file a patent application for examination. This is called "prosecuting a patent." An inventor may write and file a patent herself, but the wisest approach is to work with a patent attorney or a patent agent. People often call a specialist instead of trying to perform something complicated themselves, more likely being faster and cheaper in the long run. It's the same principle with patent applications.

To economize, Yonover and Crowe recommend inventors write much of the patent themselves. Inventors can draft the ***description*** of the invention. This part specifies the metes and bounds, what and how the invention's production and function are enabled in making a working model of the invention (the how-to aspect), and the best mode (the "preferred way of carrying out the invention").[12] This written description teaches the new art; here the inventor shares the secret sauce. The details come from the inventor's log.[13]

However, when it comes to writing the patent claims, Yonover and Crowe say inventors should leave that to an attorney or agent. **Patent claims** are statements in the application identifying novel and inventive elements of the invention. Patent claims "point out and distinctly claim the subject matter which [the applicant] regards as his or her invention," Gene Quinn, founder of IPWatchdog.com, explains.[14] Claims hold great importance for any legal consequences pertaining to an invention's patent, should disputes arise after patent issuance.

Patent claiming gradually changed over the nineteenth century. Early on, a patent's description of an invention and the patent claims together, not the claims alone, were meant to indicate the boundaries of the invention and the patent's scope. Earlier claims might only make reference to a specification on which the patent's description elaborated. This "central claiming" took a "holistic approach" to patent and claim construction.[15] By the late nineteenth century, the Patent Office, courts, and in 1870 Congress "place[d] more emphasis on the role of the claim in interpreting patent scope."[16] This "peripheral claiming" sought patent claims that would independently indicate an invention's scope. This move toward greater specificity and clarity in patent claims stands to reason, both in terms of designating the boundary lines of newly created property and as sophistication grew in the state of the art in many technological fields.

• • •

Under the patent laws of the "golden age," inventors had a **one-year grace period** before having to apply for a patent. First public disclosure of one's invention started the clock on the grace period.[17] This time allowed the inventor to discuss his or her invention with technological experts, new business partners, investors, and others. The grace period enabled an inventor to "work with others on perfecting the design and test and improve the invention."[18] This optimized the invention, strengthening the quality of the patent for success in PTO examination. These conversations didn't jeopardize the possibility of being granted a patent, didn't turn disclosures into prior art, and didn't necessarily increase the risk of an invention's theft because the patent grant decision rested on the identity of the actual inventor.

When a patent is sought, it undergoes **examination** by a PTO examiner skilled in that area of science or technology. From 1790 until 1880, those seeking a patent provided a working model of the invention, and patent models were eventually put on display. Patent models were confined in size to 12 x 12 x 12 inches.[19]

The U.S. Patent Office displayed patent models in space outfitted specifically for that purpose. Model Hall was open to the public. (Public domain. Smithsonian Institution.)

In examination, first the examiner determines if the invention is **patent-eligible subject matter**. That means determining if an invention is a process, machine, manufacture, composition of matter, or a new and useful improvement to an invention or discovery. Over time, examination practice and case law have given definition to these sorts of subject matter's patent eligibility. Naturally occurring matter such as electricity or a newly found plant species, a scientific law such as gravity, and mental steps or a process a human could perform such as mathematical calculations aren't eligible subject matter for patenting.[20] Human intervention is required.

Samuel Morse, inventor of the telegraph, and a former associate bitter with envy became embroiled in litigation over Morse's patent being infringed and countercharges that his patent was invalid, based on patent ineligibility. Henry O'Reilly asserted that one of Morse's patent claims covered a general principle of nature. The patent claim said the invention employs electromagnetism "as a means of operating or giving motion to machinery . . . [to accomplish] telegraphic communication."²¹ That Morse's claim connects the natural principles of electromagnetism with its causing machinery to operate has become an important aspect of methods and software patent eligibility.²²

Samuel Morse's telegraph receiver, patent number 6420, pictured here is housed at the Smithsonian Institution's National Museum of American History. (Public domain.)

● ● ●

If the examiner finds the invention patent-eligible, crossing this threshold moves examination to assessment of the invention on the substantive merits of patentability.²³ In the "golden age," prior art to which patents were compared was limited to inventions "known or used in the U.S. or patented or described in a printed publication anywhere . . . described in a patent or published patent application filed by another before the applicant's filing date in the U.S. [or] . . . before the applicant made the invention, it was made by another who had not abandoned, suppressed, or concealed it."²⁴ A patent must have been filed before the grace period ended. Examination looks for such qualities as "explain[ing] in specific detail how to make and use the invention covered by the patent"²⁵ and the best mode for producing the invention.

Former Chief Judge of the Court of Appeals for the Federal Circuit Randall Rader succinctly describes the difference between patent eligibility and patentability; both are elements that patent examination assesses. "Section 101 [of the patent law] requires little more for eligibility than a showing that an invention has applied natural principles to achieve a concrete purpose within the expansive categories articulated by Thomas Jefferson in 1793. Patentability, on the other hand, proceeds as a detailed claim-by-claim, feature-by-feature examination of 'the conditions and requirements of this title.'"²⁶

To be patentable, the examiner must find the invention to be novel, useful, and nonobvious. These are terms of art and quite involved in themselves.

Novel means "newness." The invention isn't identical to something already invented. That would be considered prior art. Novelty asks: Would this invention infringe another patent? That is, would the invention under examination trespass on existing intellectual property? To be novel, a patentable claimed invention must be different from prior art, inventions already in existence. Novelty is terminated by public disclosure by sale, public use, or publication.[27]

Useful means having "utility." In other words, the invention must work. It must exceed general usefulness, providing specific utility for a practical application in its subject matter. Beyond being novel, an invention must have some usefulness for something legal.[28]

Nonobvious means the claimed invention would not have been obvious to someone skilled in that art. This is based on evidence from prior art that was available before the patent was applied for. The patent examiner sees if she can find all the pieces, parts, and functionality of the invention in the prior art as a whole. Can she conclude that the combination of those elements disclosed the invention? Obviousness can sometimes be overcome by others' attempted solutions' failure to solve the problem this invention succeeds at solving. The examiner bears the burden of proving obviousness.[29]

For inventions meeting these criteria, patents are granted. During the "golden age," patent terms were fourteen years with prospective term extension of seven years. In 1861, patent terms were lengthened to seventeen years, and patent term extension was repealed.[30] For more than two centuries, U.S. patent terms began upon the date of patent grant. This makes the most sense to ensure that petitioners don't needlessly lose patent term due to office delay in completing examination.

During the "golden age," the U.S. Patent Office treated patents under examination or not granted following examination as the private property an unpatented invention actually is. Recall Chief Justice John Marshall's explanation from chapter 14 that an unpatented invention constitutes "inchoate and indefeasible property" with property rights "vested by the discovery."[31] Patent grant or not, the office treated the invention and any proprietary information about how something is made and used as secret. Protecting ungranted patents confidentially accords with the attendant private property rights in one's inventions, patented or unpatented. This "key ingredient in the U.S. patent system" made it possible for an inventor to "rework his application or use the creation as a trade secret."[32]

• • •

Once a patent is granted, what does the patent give the inventor or owner? As noted, a patent secures the exclusive right to use, make, sell, import, or license the invention or discovery. The essence of the patent right is the right to exclude others from using, making, selling, or importing the protected invention. Making, using, and selling one's patented invention is referred to as "***practicing***" one's patent.

A strict liability standard applies with patent rights. **Patent infringement** is infringement, regardless of whether the person infringing the intellectual property rights does so with knowledge the invention was protected by patent or intends to infringe a patent. There is also willful IP infringement. *Black's Law Dictionary* defines "willful" as "proceeding from a conscious motion of the will; voluntary; knowingly; deliberate."[33] Accordingly, willful patent infringement carries weightier sanctions, such as awarding punitive damages on top of actual damages (e.g., lost revenue or profits).

Patent exclusivity affords the patent owner choices among a range of options. For instance, he or she could attempt to commercialize the IP-protected invention by building a business around the invention and its IP. Another option is to sell the patent outright. Some patent owners may decide not to take immediate action on their patent rights or invention, instead letting them sit. Sometimes a patent owner will simply abandon the patent, giving up exclusivity.

Commonly, a patent owner decides to ***license*** her or his patent to a manufacturer, who will produce and sell the invention. This option results in the licensee (or licensees, if the license isn't exclusively with a single licensee) paying a licensing fee or royalties based on sales. This lets an inventor get back to inventing. Many inventors, including many of the most prolific inventors of the "golden age" such as Thomas Edison, Charles Goodyear, and Elias Howe, focus their energies on inventing—their skill set. Khan finds that inventors who licensed or sold their patents or patent rights "were the norm during the nineteenth century, and technology markets provide ample evidence that patentees who licensed or assigned their rights were typically the most productive and specialized inventors."[34] Licensors leave the making, marketing, and selling of the invention to a contract manufacturer and others better suited to the commercial endeavors involved in turning an invention into a commercial product.

• • •

Patent owners who seek to turn their inventions into commercial products often license their patents to business partners, such as by contracting with a company or companies specializing in **commercialization**, manufacturing, and related aspects of developing production, distribution, and sales chains. These partners turn the invention into a product, develop markets for the new product, and market the products through wholesale and retail outlets. Samuel Morse, for example, struck deals with several business partners—perhaps most importantly Amos Kendall. In 1845, Kendall led the commercialization efforts of "forming or licensing companies to build new telegraph lines."[35] He formed the Magnetic Telegraph Company to extend the Washington-Baltimore telegraph line to Philadelphia and New York. To some business interests, Morse granted exclusive rights to construct telegraph wires and telegraph offices for use of his telegraph in certain territories. The characteristic of patents and IP being intangible assets enables the parties involved in these activities (as opposed to invention, discovery, or creation) to pursue such economic activity. Secure ownership rights and the means of enforcing IP rights play critical roles in achieving success on the commercialization side of the coin.

Historical research shows secondary markets for selling and licensing patents began in the Antebellum Era, with "licensing . . . an essential feature of the uniquely American patent system, which secured property rights in innovation to both inventors and to the marketplace actors who commercialized the innovation."[36]

On the commercialization front, invention incentivized by property rights led to leveraging the potential and value of the inventions and patents for raising capital. Investors, financial institutions, venture capitalists, et al. were willing to risk their money based on patent rights.[37] "Patentees were able to parlay their property rights in promising inventions into part ownership in numerous companies on advantageous terms that included retention of patent rights and royalties."[38] Successful inventors with household names of yesteryear, who proliferated during America's "golden age of patenting" and whose companies bear their names, pioneered and innovated the patent licensing business model. For instance, "As [George] Westinghouse traveled restlessly among his many businesses, he continued his lifelong practice of seeking out brilliant inventor/engineers, buying their patents, and then collaborating with them to create even better industrial versions."[39]

• • •

When patent owners seek to be paid for the use of their patented invention, they assert their patent. **Patent assertion** involves the offer of a license, request for royalties or other monetary payment, or demand for compensation when patent infringement is suspected.

Though not common across the board,[40] some patent owners in these situations need to pursue litigation over their inventions and IP to enforce their rights or fend off infringers. There is a very good reason for taking civil legal action to enforce one's patent rights. There are no criminal patent statutes, and thus calling law enforcement officers to stop patent infringers or to enforce the terms of a license isn't an option. The only means available is civil lawsuit or, if a foreign-made, patent-infringing product is imported, trade laws and procedures keep those items out of the country.

The owners of foundational inventions' patents may face litigations in several judicial forums. It's fair to say the more valuable and the greater the technological leap by a patented invention, the more likely it is to end up in court. This was true for Morse and his telegraph, Singer and his sewing machine, Bell and his telephone, and the Wright Brothers and their airplane. But overall, patent litigation rates averaged 1.65 percent between 1790 and 1860, though rates vary such as during the 1840s, when patent litigation rates reached 3.6 percent.[41]

Aggressive business strategy has been more responsible for precipitating lawsuits than have weak patent rights. Likewise, patents of emerging technologies and nascent commercial sectors have been more apt to face legal squabbles.[42] For example, the 1850s saw the first "patent thicket" resulting in a "patent war," with multiple lawsuits fought over sewing machine technology.[43]

Part of patent litigation is the court's determining the scope of the patent—that is, the metes and bounds of the new property. This has generally relied on the patent claims being asserted, and the description of the invention became less central in these determinations. As more patent claims were included and individual claims have gotten longer, they have provided courts more specificity and precision with which to construe the patent's outer limits. These gradual developments over the late nineteenth century factored into judicial decisions about the **validity** of patents as derived from a patent's claims.[44] Invalidated patents or patent claims are not enforceable. Court proceedings frequently weigh patent claim validity. Decisions about validity affect a patent owner's ability to enforce broad patent scope, win greater damages awards, and inveigh on other likely patent infringers to license the patent and obtain healthier royalties.

Samuel Morse's telegraph litigation went all the way to the U.S. Supreme Court. In February 1854, the high court unanimously decided Morse was "the first and original inventor of the Telegraph," the Columbian telegraph infringed Morse's patent, and issued a "perpetual injunction" against its use.[45] The court ruled against Morse's claim of exclusivity over electromagnetism for printing characters over electrical wire, declaring that patent claim to be overbroad and, thus, invalid.[46]

The Wright Brothers litigated over their patents, as well. In *Wright v. Curtiss*, the Wrights' patent prevailed in district court in 1910; however, the Second Circuit Court of Appeals vacated the injunction and stayed the verdict. In 1913, the same district judge affirmed his ruling, declaring the Wrights' patent merited broad interpretation due to the foundational nature of wing warping, which ailerons and rudders[47] infringed. The judge issued an injunction preventing Glenn Curtiss from making or selling flying machines with infringing features. The appellate court in 1914 upheld the lower court verdict that the Wrights "may fairly be considered pioneers in the practical art of flying with heavier-than-air machines, and that the claims should have a liberal interpretation."[48] Alas, Wilbur Wright died in 1912 and didn't see the outcome.[49]

The essence of the American patent system of the "golden age" had staying power. Its fidelity to the Founders' vision and the biblical principles at the foundation of property rights and human flourishing through inventive endeavors ebbed and flowed in that period, affected by other changes in the country. The 1930s through the 1970s saw a departure in several respects from "golden age" norms. By the 1980s, high tide for American inventors and patent owners reprised.

Brilliant Golden Sky Before Sunset

The twentieth century—the American Century—witnessed some milestones by that previously obscure group of colonies that won their independence. It began with American inventors Wilbur and Orville Wright winning the global race to achieve human flight with a mechanized flying machine. The United States made a short appearance from 1917 to 1918 with the allies in Europe winning World War I. After the Roaring Twenties, Americans suffered the Great Depression of the late 1920s through the 1930s. America filled a major role in the 1940s in the victory of World War II.

Because the United States went through the Second World War unscathed by enemy bombings, its industrial base remained intact and her people confident. By contrast, much of Germany's and Japan's industries lay in ruins from heavy Allied bombardment. Soon, though, the Cold War, Korea, and Vietnam engaged the United States in espionage and combat. Profound growth of government from the 1930s forward, dramatic, escalating social change from the postwar 1940s on, and steadily growing international challenges to U.S. economic and industrial dominance in the 1960s forward took a toll on national confidence and America's strength.

Moreover, in the background during these decades, massive growth in aggressive antitrust enforcement against patents as "monopolies" that started with the Progressives ran from the 1920s to the 1970s. Meanwhile, the Supreme Court led the way in weakening patents. This combination of weakened patents and patent rights along with a heftier antitrust posture distorted the economy and shifted the innovation model away from an entrepreneurial property-rights basis. "Although [the post–World War II] period achieved important technological achievements, the postwar weak-IP regime relied on government transfers [funding] (and faltered once those transfers declined), preserved high levels of market concentration, and may have induced an organizational bias toward integrated structures for conducting innovation and commercialization activities."⁵⁰ This dynamic in government policy—against intellectual property rights and exclusivity while for aggressive antitrust enforcement—contributed to the conditions that became a drag on the country's strength and its confidence.

By 1980, America had suffered several economic shocks: Stagflation, combining negative economic growth and double-digit inflation; high taxation; high unemployment. Countries such as World War II foes Germany and Japan challenged once-dominant U.S. industrial giants. Almost a million American manufacturing jobs disappeared from a mid-1979 peak to January 1981. Beginning in the 1970s, years of trade surpluses shifted to trade deficits. Also, the OPEC oil cartel made petroleum scarce and expensive; cars and trucks lined up for rationed gasoline. Meanwhile, Iran held American diplomats hostage for more than a year.⁵¹ Unrecognized by many was that reviving the U.S. patent system's longtime strengths—incentivizing creativity with secure ownership—could be an important part of the solution to these ills.

Emerging from the troubled late 1960s and 1970s, the final two decades of the twentieth century brought the United States economic revival, victory in the Cold War with Soviet Russia, and several notable improvements to the innovation ecosystem. Among the latter were enactment and implementation of the Bayh-Dole Act, establishment of the U.S. Court of Appeals for the Federal Circuit (CAFC), and international adoption of the Trade-Related Aspects of Intellectual Property Rights (TRIPS) Agreement.⁵²

The Bayh-Dole Act

The federal government poured millions and millions of taxpayers' dollars into research and development (R&D), ramping up its R&D spending in World War II and continuing through the 1970s. By the end of the seventies, the U.S. government owned 28,000 patents on the inventions that followed—usually discovered at private companies and research universities. Yet, attempts to bring them to practical use were made on less than a paltry 5 percent of the inventions. Taxpayers paid heavily for research but gained next to no practical benefits from the many discoveries.

Government agencies made it difficult to attempt to commercialize these inventions. In general, federal agencies held title to any IP and granted only nonexclusive patent licenses, meaning the first licensee could face competition from a second licensee while trying to develop the same or a similar commercial product. This situation nullified IP's critical element of exclusivity and made the commercialization endeavor unattractive to private investors. Twenty-six different sets of rules governed licensing of a federal patent.[53]

For example, while the Department of Defense (DOD) more readily allowed contractors to take patents on their federally funded research discoveries in exchange for a nonexclusive, royalty-free license, contractors could only seek monetary damages from other government contractors who infringed their patents. DOD also liberally interpreted its license and would demand "background patents"—previously developed inventions that constitute the contractor's technical know-how and even trade secrets and proprietary data. Some defense contracts also exposed contractors to forcible disclosure of their know-how to other contractors (who may be competitors).[54]

The University and Small Business Patent Procedures Act (Public Law 96-517), or the Bayh-Dole Act named after its sponsors Sen. Birch Bayh (D-IN) and Sen. Robert Dole (R-KS), removed Washington's bureaucratic stranglehold on inventions derived from basic research under federal monies. Bayh-Dole explicitly used the property rights of the patent system to unleash the commercialization of thousands of inventions discovered at universities and federal laboratories. Enacted in 1980, Bayh-Dole gave universities and researchers exclusive ownership of any patents on their inventions. They could license those IP rights to companies and startups willing to assume the risk of attempting commercialization. Technology management was democratized, breaking the logjam in Washington.[55]

Senator Bayh lost reelection in 1980 in the Reagan presidential landslide that swept in a Republican Senate majority. The Reagan administration implemented the new Bayh-Dole law. The incoming administration hired Senator Bayh's former staff member who led the successful legislative effort, Joseph P. Allen, to help lead the law's implementation.

The 1980s saw Bayh-Dole's successful, skillful implementation in line with congressional intent, overcoming heavy bureaucratic resistance to relinquishing control over these inventions. President Reagan issued a memorandum on patent policy in 1983 extending Bayh-Dole's patent ownership to any contractor no matter its size. Senator Dole amended Bayh-Dole in 1984, removing overly prescriptive provisions related to exclusive licenses and moving statutory oversight to the Department of Commerce.

In 1986, Congress added the Federal Technology Transfer Act (Public Law 99-502) to the Stevenson-Wydler Technology Innovation Act of 1980 (Public Law 96–480), widening Bayh-Dole's reach to federal laboratories and ensuring the labs collaborate with businesses and universities. Reagan's 1987 executive order 12591 directed top officials to "encourage and facilitate collaboration among Federal laboratories, State and local governments, universities, and the private sector, particularly small business, in order to assist in the transfer of technology to the marketplace."[56]

By objective measures, Bayh-Dole has exceeded its sponsors' expectations. From 1996 to 2020, Bayh-Dole patents and licensing have added $1.9 trillion to U.S. gross industrial output, $1 trillion to U.S. gross domestic product, and more than 17,000 start-ups. It has supported 6.5 million jobs, disclosed more than 490,000 inventions, obtained more than 120,000 patents, and developed more than 200 FDA-approved drugs and vaccines.[57] This law is key to the creation of new industry sectors, including biotechnology.

The Federal Circuit

In 1982, Congress enacted the Federal Courts Improvement Act (Public Law 97-164), which established the U.S. Court of Appeals for the Federal Circuit. This law formed the Federal Circuit by merging the U.S. Court of Customs and Patent Appeals and the appellate division of the U.S. Court of Claims.[58] The Federal Circuit received sole, nationwide jurisdiction over appeals cases of several subjects, including patents and federal registration of trademarks.[59]

This new judicial forum's creation was prompted by "regional appeals courts [having] considerable leeway to interpret patent law as they saw fit."[60] That discretion led to wide variation in circuit courts' application and interpretation of patent law. Judges, many of whom had little background in patent law, decided cases effectively determining whether a patent held value. Further, technological complexity could be challenging for jurists less familiar with it or who rarely heard patent appeals. The same patent could be deemed of great economic value in one circuit and of much less value in another appellate court. The validity of patents on appeal dove in the years ahead of the CAFC's creation.[61]

This concern of discordant patent rulings across the circuits, along with the Supreme Court's unlikely resolution of circuit splits over patents, raised doubts about patent reliability. The resulting "uncertainty had undermined the patent system, and a designated national appeals court could bring about consistency and promote the innovation necessary for American businesses to compete in an increasingly competitive global environment" the United States found itself in while suffering bleak economic conditions.[62] The appeals courts contributed to weakening patents. Greater consistency of patent law was needed.

The CAFC acted to address Congress's intentions for the new court. In its first case, all the Federal Circuit judges together adopted the rulings of the two merged courts that constituted the Federal Circuit as its precedents. This move immediately gave the sole appellate court uniform rulings to follow, unless they encountered conflicting decisions—with reversal then only with all CAFC judges sitting *en banc*.[63] This "resolved all of the regional circuit splits concerning patent law in one fell swoop and created a mechanism for resolving panel disputes ensuring uniformity to the law and certainty in the patent system."[64]

Additionally, CAFC ruled patent claim construction is a matter of law and therefore a judge's decision, rather than a question of fact that juries determine. This 1995 holding in *Markman v. Westview Instruments*[65] further brought certainty and predictability to patent jurisprudence.[66] The Federal Circuit also decided patent eligibility in 1998 in *State Street Bank & Trust Co. v. Signature Financial Group*.[67] In *State Street*, the Federal Circuit ruled a mutual fund investment management software program was eligible subject matter for patenting. CAFC upheld the validity of patenting a business method for computing, based on its tangible, concrete, useful results, leading to increased business-method patent filings.[68]

The Federal Circuit's steadying effect was generally widespread. For the semiconductor sector, CAFC appears to have contributed positively. Invented in the late 1950s, semiconductors (also known as microchips or integrated circuits) followed the transistor, invented in 1947 and part of the circuits on microchips. Patents of semiconductors issued yearly sat at about 1,000 U.S. patents per year from the mid-1970s to the mid-1980s. Within a decade of the Federal Circuit's creation in 1982, semiconductor patents rose to about 3,000 granted annually in 1992. Patents on these inventions rocketed to about 10,000 a year in 2000, doubling that annual number by 2017.[69]

This improvement in a capital-intensive, critical technological sector correlates with the Federal Circuit's formation. Some researchers attribute this to the CAFC's role in making U.S. patent rights stronger. People in the industry attributed patenting's rise, after the Federal Circuit came into being, as a means to protect themselves. Semiconductor firms faced more litigation seeking to halt their commercial efforts with an injunction that stopped production. "A [semiconductor] firm with a large patent portfolio could countersue, thus making lawsuits less likely."[70]

Thus, the Federal Circuit had a near-immediate, salutary effect on patent jurisprudence. It increased uniformity and certainty.

The TRIPS Agreement

The United States pressed for international intellectual property protections for years, but the outlook remained dim. In 1983, President Reagan formed a Commission on Industrial Competitiveness. Part of its mission was to examine the role of IP in advancing U.S. industrial competitiveness through both domestic and international IP reforms. A committee of the Young Commission—so called after its chairman, John A. Young, president of Hewlett-Packard—addressed this. In 1984, the committee wrote:

> Disillusionment with intellectual property deliberations in WIPO and UNCTAD[71] has prompted a number of developed countries, most notably the United States, to investigate the possibility of seeking relief through the General Agreement on Tariffs and

Trade (GATT), . . . the primary international agreement governing trade between nations.

Efforts to date to address intellectual property rights in GATT have focused on an anticounterfeiting code, since counterfeiting most directly and significantly affects trade flows. The United States initially suggested that rules were needed to cover the counterfeiting of all forms of intellectual property, including patents, copyrights, trademarks, and industrial designs, but there was no international support for such broadly defined intellectual property coverage under GATT. Today, discussion over a draft agreement within GATT is limited to trademarks. The developing countries, in particular, have resisted efforts to discuss intellectual property issues in GATT, arguing that intellectual property issues belong in WIPO exclusively.[72]

Foreign governments' signing multilateral agreements is no guarantee a signatory will live up to the terms it has pledged. Various nations' policies and practices (then and now) enable expropriation, compulsory licensing, and patent forfeiture; provide inadequate IP protections, weak judicial enforcement mechanisms, and means that facilitate taking IP and denying the IP rights of innovators from other countries.[73]

Still, the Young Commission recommended advocating a universal international IP agreement to help remedy the anti-IP posture. "The United States should consider the feasibility and scope of a universal agreement on the protection of all forms of intellectual property rights . . . to provide effective, comprehensive international protection of intellectual property rights . . ."[74]

In 1995, the United States prevailed in having IP protections included in international trade. The Trade-Related Aspects of Intellectual Property Rights Agreement came under the purview of the World Trade Organization (WTO) in the WTO's organizing document. Key aspects of TRIPS require countries to afford foreign IP owners the same treatment as its citizens, called "national treatment;" a term of agreement a nation gives to one WTO member must also be available to another member nation.[75] TRIPS requires all WTO member nations to adopt IP statutes protecting IP and IP ownership, ensure judicial systems by which IP rights can be adequately enforced, and pursue domestic IP thieves and counterfeiters.[76] For example, TRIPS sets a minimum for domestic patent terms of twenty years from date of filing. The WTO adjudicates dispute resolutions under TRIPS, with authority to penalize nations found in violation of the agreement.

The TRIPS Agreement was intended to strengthen IP rights in the context of international trade. Practical enforcement in many foreign nations of IP rights against pirates, counterfeiters, and other infringers had been virtually impossible. The United States aimed for TRIPS to lift other countries' IP regimes and legal rights more closely to resemble the "gold standard" American IP system.

* * *

Engineer and inventor Susie Armstrong says having good patent policy "makes a huge difference." Armstrong, who at Qualcomm was central to the invention and commercialization of technology that enabled web-surfing on a cellular phone in 1997, believes "having a strong patent system has . . . pretty much driven American competitiveness . . . since day one" across types of technology and useful arts.[77]

In important ways, the U.S. patent system as it existed from 1836 until the end of the twentieth century sat at the center of America's innovation ecosystem. "Patent law kept the character that has defined most of its history since 1836: adaptive but essentially stable in its framework" (internal citation omitted).[78] U.S. patenting along with the facets discussed above—a framework for assessing patents expertly, impartially, and confidentially; secure, enforceable private property rights in patents and IP; recognition of patents as valuable economic assets; a court system that operates by the rule of law, impartiality, fairness, and due process in recognition that patents secure property rights; and granting U.S. patents based on the objective merits of an invention, as newly created property, without regard to the status or characteristics of its inventor—constituted a "virtuous circle."

America's virtuous circle delivered world-changing benefits to individual human beings, far-away parts of the country, the United States as a nation and as an economy, and to the ends of the earth. No other patent or intellectual property system—or national innovation ecosystem—has embodied the biblical principles of creativity and ownership as did the United States during the "golden age of patenting."

Chapter 16
A Divide Between One's Creations and Secure Ownership

"How the mighty have fallen!"
— 2 Samuel 1:19b

"Those who cannot remember the past are condemned to repeat it."
— George Santyana[1]

"Boris Teksler, Apple's former patent chief, observes that 'efficient infringement,' where the benefits outweigh the legal costs of defending against a suit, could almost be viewed as a 'fiduciary responsibility,' at least for cash-rich firms that can afford to litigate without end."
— The Economist[2]

The "virtuous circle" of U.S. patenting 3.0, the "golden age of patenting" from 1836 toward the end of the twentieth century, generally reflected biblical principles of creativity and ownership. Though imperfect as is any human endeavor, this model—or innovation ecosystem—worked extraordinarily well for many, many years.

That said, many evolutionary changes in the United States and the world as well as certain revolutionary changes occurred during that period—some positive, others quite the opposite. Many of them affected the innovation system, directly or indirectly. For instance, a positive effect during the Civil War years came in the form of a spike in innovation. More than 16,000 U.S. patents were issued from 1861 to 1865, and the Confederate patent office granted 266 patents.[3] Another effect: The war effort boosted industrialization on both sides of the Mason-Dixon Line, though mainly in the North.

Late nineteenth-century industrial growth, often involving the commercialization of patented inventions, created great wealth, brought the nation and the world remarkable new technologies, and provided jobs, labor-saving devices, and new diversions (e.g., movies, sound and musical recordings, telephones).

George Westinghouse and Nikola Tesla together developed a hydroelectric plant. Their system generated, transmitted, converted, and distributed AC electric current. This was "disruptive" technology—the sort of invention that replaced candles and oil-burning lamps with a new energy source, soon powering everything from homes to factories. By the turn of the century, Wilbur and Orville Wright stood on the cusp of achieving human flight.

A Change of Heart

Beneath all this, human hearts, souls, and minds changed over the nineteenth and twentieth centuries. The beneficial fires of the Gospel at the center of the First Great Awakening cooled to smoldering embers over the next two centuries. Christian intellectual Francis Schaeffer points to philosophical shifts during the eighteenth century that weakened Reformation thought.[4] Enlightenment philosophers Jean Jacques Rousseau and Immanuel Kant, among others, loosened the prevailing view of nature and grace. Determinism became applied to humans, absolute freedom replaced grace, and hope was lost for understanding reality with a unified field of knowledge.[5]

What Schaeffer calls "the line of despair" reflects the embrace of these shifts in thought. In the nineteenth century, Georg Hegel's philosophy broke from "the concept of a rational, unified field of knowledge." Schaeffer observes, "The distinctive mark of the twentieth-century intellectual and cultural climate . . . [lies] in the unifying concept . . . of a divided field of knowledge. Whether the symbols to express this are those of painting, poetry or theology are incidental. The vital question is . . . the concept of truth and the method of attaining truth."[6] Hegel's concept of higher truth derived from synthesis of opposite theses, or philosophical propositions.[7]

Perception of truth influences human beings' understanding of the Creator, the objective universe, and all facets of God's creation. This includes natural phenomena such as scientific and mathematical laws; materials such as metals, plants, and their respective properties; and humanity's physical, mental, and spiritual attributes. The beginning of the "golden age of patenting" coincided with prevailing beliefs about truth, reality, etc. that, among much of society, were generally the Judeo-Christian worldview. Society's hewing closely to the Creator's and creation's reality enabled both Christian and non-Christian inventors and scientists, under Common Grace and the Creation Mandate, to unlock levels of understanding and apply their discoveries to practical use.

Gradually over the past two centuries, a Reformation understanding of a unified, rational field of knowledge under the God of the universe changed to a concept of truth holding to a divided understanding of knowledge without the Creator. Schaeffer's examination of these philosophical shifts illuminates how they filter through philosophy, human intellectual and creative endeavors, and theology.[8]

Consider the Baroque music (1600–1750) of composers such as Bach, Handel, and Vivaldi, its diversity within unity, contrapuntal dissonance and resolution. The Classical (1750–1820; Beethoven, Mozart, Haydn) and Romantic (1820–1900; Chopin, Brahms, Tchaikovsky) eras that followed exhibited unique developments characterizing each era. Still, form and manner retained objective values connected to truth, beauty, and goodness.

Now contrast Modern era music (1900–present; early ones included Stravinsky, Ravel, Holst). French composer Claude "Debussy (1862–1918) is the doorway into the field of modern music [for the unifying concept of a divided field of knowledge]."[9] Debussy was "the instigator of the entire 'modern' movement" whose music shocked with "harmonic audacities" and other avant-garde techniques.[10] Modern composers have shunned musical conventions, constructing "a place of complete free reign . . . a place for the ultimate [musical] experimentation."[11]

Twentieth-century "classical" music got uglier and more grating on the ears, and still hasn't turned a corner toward the true, good, and beautiful.

Anything that sounds terrible is lauded by the critics. We reached a point where compositions of beauty, harmony, and grace are treated disgracefully. It happened to Gian-Carlo Menotti in the mid-twentieth century. His Christmas opera *Amahl and the Night Visitors* pleased audiences, but the "professionals" shunned and derided.[12] Today, the lovely compositions of an extraordinary talent, child prodigy Alma Deutscher, aren't liked by critics because it *isn't ugly*. In fact, it's truly beautiful and harmonic. Deutscher composes lovely, pleasing, refreshing musical works audiences love because she believes "music should be beautiful."[13] Deutscher fashions her music in a manner more like that of Mozart and other truly great composers whose music stands the test of time than that of the shock value–laden twentieth- and twenty-first-century avant garde noise.

The progression of philosophy untethered from the Bible has spread to many areas of life and human thought. It has "rooted [modern science] in naturalistic philosophy"[14] and entered law, politics, and much more. The effects on humanity's view of creativity is akin to the contrasts of Baroque and Modern (or Postmodern) music; the true, good, and beautiful have been replaced by the false, evil, and ugly—in the name of human autonomy. This worldview permeates humanity's view of ownership, infiltrating humanity's views of invention and property rights over one's creative works. It fundamentally breaks with the American Founding's biblically informed perspective on human nature and central tenets such as property-rights principles.

It Gets Worse

Economic and social change tied to industrialization and inventions affected peoples' views of those developments. Companies, such as Westinghouse, Edison Electric (changed to General Electric in a merger transaction), Bell Telephone, and Eastman Kodak, were founded by prominent inventor-entrepreneurs. In the late nineteenth century, large corporations combining into "monopolies" in capital-intensive sectors attracted public attention and, increasingly, government scrutiny. People generally marveled at inventors' amazing achievements. But the financiers and the tycoons in those industries, such as Jay Gould,[15] and firms organized as "trusts,"[16] such as John D. Rockefeller's Standard Oil, began to color public sentiment. Some people had misgivings about "big business."[17]

Beginning near the close of the nineteenth century, Congress enacted landmark antitrust laws empowering the government to regulate certain competition-related business practices and to pursue "trust busting"[18]—countering consolidation of oil refineries, railroads, steel, and other large industrial firms. The Sherman Act became law in 1890, and the Clayton Act and the Federal Trade Commission Act (FTC Act) both were enacted in 1914. Sherman aims to halt restraint of trade pursued through combining companies and monopolizing. Clayton targets actions that reduce competition or create a monopoly. The FTC Act creates an administrative agency to enforce antitrust laws against unfairly anticompetitive conduct.[19]

At the turn of the century, some business executives were taken in and were joined by academics, journalists, and other cheerleaders in popularizing the science of administration. Morris says this fad, which lasted from the late nineteenth- through the early twentieth-century decades (with vestiges hanging on through the twentieth century), was the extensive application of scientific methods and measurements to complex businesses. The same enthusiasm spilled over into other areas, such as sociology, the American Society of Mechanical Engineers, and John Dewey's educational philosophy.[20] "The cult of the expert was born," Morris writes.[21]

In the 1920s, the American economy was booming. By the end of the decade, the good-times bubble went "pop." The stock market crashes of October 1929 got America's attention. Subsequent bank failures in the early 1930s due to easy credit led to the Great Depression. Credit tightened and wealth got wiped out, accompanied by widespread economic disruption and unemployment throughout the nation.[22]

In 1932, Americans elected New York Gov. Franklin D. Roosevelt as president. Roosevelt's First Hundred Days, with the help of Congress, swept in "a body of far-reaching and direction-altering legislation," policies collectively known as the New Deal.[23]

FDR's New Deal, Carson says, rested on four key ideological pillars. *Populism* advocated money untied from gold's value, easing credit by making paper currency a tool for feeding inflation. *Progressivism* pushed activist government driven by the Constitution's Article II executive branch, displacing the Article I legislative branch in making law. Progressive "reform" meant "government regulation of big businesses rather than breaking them up by applying antitrust laws."[24]

U.S. mobilization for World War I influenced many of FDR's officials, who thought the model of wartime collective efforts on military and civilian fronts, bolstered by bellicose language, could successfully be applied to counter a national economic crisis. Capturing the imagination of many New Dealers was the belief that a *centralized economy*, in which government functionaries plan and manage the nation's economic system, could outperform free enterprise, which they derided as wasteful and disorganized.[25] Together, these two factors represent an "-ism": collectivism.[26]

Taken together, Carson's four ideological pillars of populism, progressivism, and World War I mobilization's influence on New Dealers' faith in central planning reflect the impact of the turn-of-the-century's "cult of the expert" at its apex thirty years earlier. A generation steeped in the "science of administration" and the superior judgment of "experts" naturally incorporated this philosophical mix into the crisis at hand.

Thus, humanity let up holding fast to biblical truths and placed faith in human institutions, especially government. This had the steady effect of degrading foundational elements of the innovation ecosystem of the golden age of patenting. Coupled with progressivism's dream of a ruling class of allegedly "nonpartisan experts" to micromanage the economy and regulate the private sector, these trends added the fuel of bureaucratic autonomy to the fire of spreading despair. It rejected the Founders' biblically steeped republican government and Lincoln's fuel of interest to the fire of genius innovation model.

This turn amounted to "exchang[ing] the truth about God for a lie and worship[ing] and serv[ing] the creature rather than the Creator . . ." (Rom. 1:25). No longer were God and human beings regarded as outside what Schaeffer calls "the machine"—the parts of creation that lack rationality and a soul. This shift in worldview devalued attributes God has communicated to his highest creature, including creativity and ownership, disavowing property rights and inherent ownership.

* * *

As with any new laws, courts hearing antitrust cases interpreted the law in specific cases and fact patterns. Judge Robert Bork notes exclusive terms in contracts "are forms of vertical integration" that antitrust law has treated "severely."[27] The philosophical trends filtering into other areas of society in the nineteenth and twentieth centuries also reached antitrust. Antitrust enforcers have regarded exclusivity with skepticism because of the potential for monopoly and diminished competition. For a lengthy portion of the golden age of U.S. patenting, antitrust was wielded against patent owners, due to the vertical restraints patent exclusivity provides.[28]

Bork cites four trends in antitrust law, echoing our earlier discussion of fool's gold replacing societal blessings from God's created order:

> (1) a movement away from political decision by democratic processes toward political choices by courts; (2) a movement away from the ideal of free markets toward the ideal of regulated markets; (3) a tendency to be concerned with group welfare rather than general welfare; and (4) a movement away from the ideal of liberty and reward according to merit toward an ideal of equality of outcome and reward according to status. Common to all of these movements is an anticapitalist and authoritarian ethos. It is that spirit which creates these tendencies, and it is their success which reinforces that spirit.[29]

Given this course of human events, antitrust specialists tend to take issue with IP exclusivity. During the phase of American inventing and patenting's golden age that coincided with the Depression, the New Deal, and thereafter, an IP rights-skeptical antitrust approach dramatically tilted the scales of justice against patent owners' property rights and thwarted the dynamic competition innovation produces. The foundations of reliable ownership of one's creative works in the context of secure private property rights crumbled as antitrust acid dripped on IP exclusivity. The Patent Bargain fell out of balance.

The 1930s through 1970s witnessed dramatic change where "courts and regulators were largely unsympathetic toward patents and expansively interpreted antitrust constraints on patent licensing."[30] Property rights in patents were weakened, beginning in the 1930s, as courts increasingly invalidated patents in infringement cases over the next several decades. A 1931 Supreme Court ruling created a patent "scope" doctrine, which gave infringers a pliable defense of "patent misuse" in court.

Antitrust laws were enforced against IP owners and patent licensees who asserted their exclusive rights. In 1942, the high court indicated that *patent infringers could use the scope doctrine to sue patent owners on antitrust grounds.* Jonathan Barnett notes this development exposed patent owners to monetary damages, "rais[ing] the possibility that the patent owner can be held liable to the alleged infringer, which obviously has a deterrent effect on patent enforcement actions."[31] He further explains the combination of the "patent misuse doctrine" and antitrust common law regarding vertical contracts approach "per se[32] prohibitions of a large menu of [patent] licensing terms, even without evidence of market power or anticompetitive effects."[33] Seeking to enforce patent rights against infringers could at that point trigger antitrust liability.

This *Alice in Wonderland*–like situation became surreal, from the perspective of a biblical ethos and Western property rights. Competition agencies sought compulsory licenses, whereby the government or other parties may infringe IP-protected creations, using them without permission, perhaps for an often low, in many cases zero, royalty. Starting in 1938, antitrust enforcers secured compulsory licensing of patents primarily by consent decree or else through litigation in court.[34] The example of the U.S. military pressing the Wright Brothers, Curtiss, and other aviation innovators to pool and cross-license their patents set a damaging precedent. Twenty years later, compulsory licensing came under the threat of the antitrust cudgel. By then, courts were unfavorable toward patents.[35]

Other changes coincided with antitrust's anti-IP bent to entrench the weak-IP, robust-antitrust regime. First, compulsory licensing orders as antitrust remedies became most frequently issued between 1945 and 1958.[36] Also, during the decades of the war and afterward, the U.S. government poured money into research and development. This funding most often went to universities and large corporations. This "tool for market re-engineering" enabled the government to pick winners and losers "to institute an ideal of atomistic competition, sometimes to the detriment of firms making pioneering technological advances in a particular market."[37] Federal R&D funding contracts contained terms that, for instance, limited patenting of associated inventions, provided for compulsory licensing, or imposed limits on enforcing IP rights. The level of federal funding was so great, it pushed aside private R&D investment.[38]

The forty-plus-year weak-patents, robust-antitrust regime displayed the failures of central planning. As Barnett documents, this government interference in the marketplace favored a few large corporations, disfavoring small- and medium-sized, innovative businesses that might inject dynamic competition around their inventions—i.e., the government from the New Deal through the 1970s actually boosted market concentration, sustaining an oligopoly of a handful of powerful corporations dependent on federal largesse. It substituted taxpayer R&D money, displacing private investment. This eventually costed U.S. competitiveness and weakened the U.S. economy. It teed up the wasteful situation the 1980 Bayh-Dole Act corrected— lots of new knowledge, but negligible practical benefit due to constrained IP exclusivity.[39] Bayh-Dole was aided by the judiciary's adoption of Bork's objective, evidence-based Consumer Welfare Standard in antitrust law.[40]

Condemned to Repeat History

The slide America suffered by the 1970s was due in large part to policies of the New Deal and the postwar era, including the diminution of patents and IP ownership rights and the aggressive, misguided antitrust regime of the time. When the 1980s arrived and strengthened property rights and economic freedom, it felt like, and indeed it was, "morning in America."[41]

Over time, memories faded of just how bad things had been those decades when patents were weakened, ownership rights reduced, and creativity and invention subjected to government expropriation at home and abroad. Those with other ideas—returning to weak patents and strong antitrust—never went away or heeded the lessons history held.

Thus, in recent years, the prospect of the United States' repeating a dark era of our history has steadily gained footing. Global commerce expanded, with the United States yielding ground to other nations' trade demands in the process. For instance, in getting the 1995 TRIPS Agreement, the United States gave up the right to apply unilateral trade sanctions in IP disputes against other World Trade Organization (WTO) nations. Choate calls the loss of this power "intangible."[42]

Transnational companies have a much greater need for access to international markets, including for their IP, than small and medium-sized American enterprises. Some corporations are legacy firms benefiting from the World War II–postwar economy; they relied on U.S. government funding and contracts for R&D and innovated under the distorted, weak-IP, strong-antitrust regime. "While their global agenda was to harmonize the IP laws and procedures of the developing world and bring them up to the U.S. standard, their domestic goal was to harmonize the U.S. patent standard down to Europe's level."[43]

The first part, which TRIPS promoted, is laudable. But the sacrifice of key facets of the unique, superior American patent system is misguided and regrettable. It certainly breaks faith with biblically informed property-rights principles behind American patenting 3.0.

The changes international companies envisioned for the U.S. patent system to achieve global "harmonization" have floated around Washington for years, appearing in various commissions' recommendations.[44] These include reducing the length of patent terms and starting the term based on date of patent filing instead of grant; publishing patent applications just months after a patent has been filed; establishing adversarial proceedings in the U.S. Patent and Trademark Office (PTO) in which third parties can challenge patent validity (similar to forums in weaker patent systems); switching from first-to-invent to first-to-file as the basis for determining to whom a patent is granted; creating prior user rights, which allow patent infringers to avoid paying royalties if they used something akin to the patented invention; and privatizing the PTO.[45]

Having a more uniform set of patent laws internationally is understandable from the perspective of global corporations, all things being equal. However, it makes little sense, from the standpoint of patent law quality, secure private property rights, or incentive for invention, to eliminate valuable assets in a country's patent laws—namely, those laws preserving and protecting private property rights in one's inventions.

Nor are all things equal: Many nations have little to no respect for property rights, the dignity of individual human beings, or the inherent right to own what one creates. Thus, "harmonization" on such an uneven field is foolish and shortsighted, especially by dumbing down the globally superior patent system. "Their code word is harmonization," Phyllis Schlafly warns. "It's a betrayal of American inventors to harmonize down to inferior foreign practices; we should encourage them to harmonize up to our proven system."[46]

A Divide Between One's Creations and Secure Ownership | 219

● ● ●

Since the 1990s, the knives have been out, aimed at the heart of the American patent system. Not only foreign adversaries and corporate globalists have worked to weaken the U.S. patent system. Congress, presidential administrations, and courts have inflicted harm on American patenting. While some facets of the contemporary U.S. patent regime remain from the 3.0 version, an argument can be made that America is well on the way, if not already shifted to, American patenting 4.0—representing regress, not progress.

What has occurred to weaken the U.S. patent system—and the private property rights–based patent incentives for advancing the progress of science and useful arts through individual creativity, ingenuity, and secure ownership rights? Legislation, executive actions, and court rulings by officials of both parties have harmed IP-related flourishing.[47]

Congress, at the Clinton administration's behest, shortened the U.S. patent term. Legislation to implement the General Agreement on Tariffs and Trade (GATT) that created the World Trade Organization (WTO) and enacted the TRIPS Agreement passed Congress in the 1994 lame-duck session. This legislation set U.S. patent terms at twenty years from when a patent is filed. Patents had provided seventeen years of exclusivity from when a patent was granted. Thus, the TRIPS minimum patent term became the U.S. maximum. Under the traditional American model, patent rights guaranteed a reliable period, unaffected by protracted PTO examination.

Congress and the Clinton administration required the PTO to publish patent applications undergoing examination eighteen months after a patent is filed.[48] The 1999 American Inventors Protection Act enacted the mandatory publication requirement and provided certain prior user rights. Japan and China lobbied for these and other changes. Premature public disclosure makes an invention prior art. Inventors lose their property rights to their inventions because publishing applications denies them the option of keeping and using their invention as a trade secret. The silver lining is a narrow exemption from automatic publication. PTO may not prematurely publish the pending application of an inventor who foregoes patenting in foreign nations.[49]

• • •

The most far-reaching patent legislation Congress has recently enacted is the 2011 America Invents Act (AIA).[50] The AIA's contents stem from the "global harmonization" policy agenda discussed above, its debate continuing over several Congresses through the 1990s and 2000s. AIA wrote into patent law administrative proceedings inside PTO through which patent validity may be challenged in adversarial quasi-judicial tribunals. The notorious *inter partes* review (IPR) proceeding, post grant review, and Covered Business Methods (CBM) program[51] in the Patent Trial and Appeal Board (PTAB) all come from the AIA. PTAB is so offensive, it has driven inventors to burn patents in protest.[52]

PTAB was promoted during AIA debate as a faster, cheaper, more efficient alternative to litigating in federal court to resolve patent validity disputes. Its advocates asserted PTAB would give patent challengers "one bite at the apple." Proponents claimed "weak patents" and a "litigation explosion" driven by "patent trolls" weighed on the patent system and innovators. Those claims were little more than spin.[53]

Created by Congress and the Obama administration, PTAB's administrative invalidation of issued patents deprives owners of property rights. All property claims in the United States have guaranteed Fifth and Seventh Amendment protections, such as the right to a jury trial, de novo review, the clear and convincing evidence standard of proof, and those suing must have standing, a dog in the fight. When the government takes away property, the Constitution requires it to provide the property owner, including patent owners, just compensation.

Yet, PTAB proceedings allow anyone, including a patent owners' competitors, to pursue patent validity challenges. Repeated PTAB validity challenges can come anytime during a patent term, denying quiet title.[54] Obnoxiously, a federal court ruling on the same patent claims does not keep PTAB from overruling the Article III court.[55] If a patent owner prevails in a PTAB proceeding, the only thing she wins is keeping her patent. And no compensation is paid if PTAB invalidates one's IP.

Moreover, the Obama PTO exercised broad bureaucratic discretion in setting up PTAB.[56] Thus, PTAB allows itself to ignore due process, fairness, procedural norms, and judicial ethics. It ignores the provision of patent law that requires presuming an issued patent's validity. While federal courts apply the "clear and convincing evidence" standard of proof to overcome the presumption of validity, PTAB uses the lower "preponderance of the evidence" standard.[57]

Gene Quinn notes PTAB's "rules of procedure [are] . . . slanted so noticeably toward the challenger and against the property owner who is supposed to own a patent that is presumed valid . . . [while] PTAB interpretations of the rules and PTAB decisions lead to such obvious procedural unfairness that even a first year law student would notice the obvious lack of due process."[58] PTAB also cooks the books, known as "IPR [inter partes review] gang tackling," producing misleading statistics that muddle the facts.[59] PTAB grants administrative patent challenges in parallel (multiple proceedings at the same time against the same patent at PTAB or simultaneously with judicial litigation), serial (repeated challenges against the same patent), and against a patent a federal court or the ITC has ruled valid.

Comparing PTAB to federal court on patent rulings, PTAB invalidates far more patent claims than do Article III courts. PTAB invalidates at least one claim up to all patent claims in 84 percent of patents reviewed.[60] The Congressional Research Service, nonpartisan analysts at the Library of Congress, affirms "PTAB cancels at least one patent claim in over 80% of its final written decisions[;] . . . that number is accurate . . ."[61] By contrast, federal district courts determine patents to be invalid about half that rate. One study puts federal court invalidation at 42.6 percent.[62] Part of the problem is:

> PTAB routinely prohibits amendments [to patent claims], which means it routinely invalidates patent rights that could be revised and preserved by accommodating new information. This explains the inordinately high invalidation rates at the PTAB in comparison with federal courts—which have long interpreted patents under the same rules that apply to all legal documents securing secure private rights of contract or property.[63]

Also, administrative patent judges (APJs) are federal employees, not actual judges. APJs use "the slanted playing field of the PTAB . . . resulting in the invalidation of patents upheld in district courts."[64] APJs are unbound by any code of ethics satisfactorily similar to that to which judicial branch jurists are accountable. This leaves innovators, such as Centripetal Networks[65] and VLSI, exposed to patent validity challenges by deep-pocketed opponents, seriously threatening their patents.[66] By contrast, Article III judges are presidentially nominated, Senate-confirmed to lifetime terms.

From 2018 through 2020, PTO Director Andrei Iancu made several administrative reforms that began to improve matters for inventors and patent owners.[67] Iancu's efforts included providing examiners guidance for determining patent eligibility, changing the legal standard PTAB applies when construing patent claims to the same standard Article III courts and the International Trade Commission use in determining patent validity, and a joint policy statement with the Justice Department and the National Institute of Standards and Technology (NIST) securing standard-essential patent (SEP) rights. However, the Biden administration weakened many of those reforms, even applying an antitrust lens to diminish patents and patent rights, such as pharmaceutical patents.[68]

• • •

Also, the AIA dropped the American practice of reserving patent grants for the first, actual inventor—a blow to the sanctity of private property rights. Now, the first petitioner to file a patent receives the IP rights. And Congress eliminated the one-year grace period after first sale or public disclosure, substituting a "provisional" patent that aligns with first-to-file instead of ownership rights of the true inventor.

Congress extended "prior art" to the ends of the earth in the AIA. Previously, U.S. searches for prior art were generally limited to prior art in the United States. After all, U.S. patents only have force in the U.S.A. Further, the AIA empowers third parties from anywhere to submit prior art, which now becomes part of the PTO's record.[69] This expansion unnecessarily exposes U.S. patents to rejection in examinations and invalidation once granted.

The AIA expanded "prior user rights." More than a year's unauthorized use of technology like a patented invention frees patent infringers from liability for royalties. The Court of Appeals for the Federal Circuit handed prior users a defense against infringement claims in *State Street Bank & Trust Co. v. Signature Financial Group* in 1999.[70] Congress expanded

The U.S. Patent and Trademark Office is headquartered in Alexandria, Virginia. The building pictured has been the office's home since 2003. (Photograph by MacKenzie R. Edwards.)

these rights.[71] AIA prior user rights "create an incentive for prior users to keep their technology secret by eliminating any liability [for infringement]."[72] That is, prior user rights make trade secrets attractive rather than disclosure-for-exclusive-rights.

Prior user rights carry great cost at odds with ownership and exclusivity:

> Prior user rights . . . undermine [the] exclusivity principle. With prior user rights, the patentee may no longer enjoy the limited monopoly afforded to him as a result of obtaining a patent of his invention. Prior user rights allow for the secret use of the subject matter without a license and free from liability to the patentee. The patentee loses his right to exclude guaranteed by § 271 [of Title 35] at no cost to the prior user. Therefore, the prior user weakens the rights afforded to the patentee, which violates the [Patent Bargain's] coveted reward for disclosure.[73]

This clearly undermines property rights.[74]

• • •

In antitrust, too, backsliding has occurred toward the IP-skeptical days of the 1930s to 1970s. The Obama administration's Federal Trade Commission launched controversial antitrust litigation against wireless technology innovator Qualcomm just days before the administration expired. The FTC's move targeted Qualcomm's SEP licensing practices. After nearly four years of litigation, the Ninth Circuit Court of Appeals handed down its ruling: "A three-judge appellate panel overturned the trial court's errant ruling, giving the FTC a comeuppance in its antitrust suit against Qualcomm . . ."[75] In that litigation, the FTC assaulted the exercise of IP rights while the U.S. Departments of Justice (DOJ), Energy, and Defense sided with Qualcomm.[76]

A brief rebalancing occurred under DOJ Assistant Attorney General for Antitrust Makan Delrahim. The first patent attorney to head the Antitrust Division, Delrahim structured and pursued a "symmetric" approach to antitrust when IP is concerned.[77] His New Madison Approach hinged on the dynamic competition patented innovation creates, adjusting antitrust enforcement accordingly. The New Madison legal framework, which applied in the *FTC v. Qualcomm* case,[78] was taking root in cases when the Trump administration and Delrahim left office.[79]

The Biden administration[80] and populists of both political parties[81] have pursued a decidedly weak-IP, aggressive-antitrust policy agenda.[82] Warnings such as one in *National Review* ("Many of the underlying abuses associated with the [trusts of the] Gilded Age were enabled by an unhealthy relationship between government and capital that distorted the operation of free markets.")[83] regarding misplaced antitrust focus have gone largely unheeded.[84] This has left private enterprises to guess the disposition of competition agencies toward routine business decisions and practices, often carrying expensive legal costs.

For IP-centric companies—the types for whom SEPs and IP exclusivity are paramount, such as InterDigital and Qualcomm—tilting antitrust against and weakening IP rights of innovators threatens R&D funding, the ability to assert their patents against free-riding or uncooperative implementers, and thus their competitiveness against Chinese and other nations' government-owned and -subsidized competitors.[85]

At their best, Congress and the executive branch serve the checks-and-balances functions the Founders intended. This should drive them to reasonable compromise and constrain broad, overreaching laws and policies. However, in the patent-IP policy arena, these branches have done more to loosen than secure America's intellectual property and associated rights to enforce IP exclusivity.

Disorder in the Court

Courts share responsibility with Congress and executive agencies for the diminished state of our patent system. The muddled status of patent-eligible subject matter doctrine, reduced access to injunctive relief, and questions about the very essence of patents as private property have thrown U.S. patenting into confusion.

Patent Eligibility

The Supreme Court's 1980 ruling in *Diamond v. Chakrabarty*[86] may be considered a landmark in patent eligibility doctrine. The court found a man-made, living microorganism eligible subject matter for patenting. The court said section 101 of the patent law regards "anything under the sun that is made by man" as patent eligible. *Chakrabarty* seeded the biotechnology revolution, sprouting new benefits for humankind. This and similar rulings strengthened IP rights in the United States. If *Chakrabarty* marked a high point for patent eligibility in the modern era, more recent judicial renderings mark a nadir (assuming we've reached bottom destabilizing and weakening patent rights), foreclosing what should be eligible for patent protection.

Four landmark 101 cases define the crux of the chaos.[87] In its 2010 *Bilski v. Kappos* decision,[88] the Supreme Court clouded the patent eligibility of business methods. *Bilski* casted uncertainty on whether the process an invention enables is an "abstract idea" and thus disqualified. This limited the patent eligibility of certain computer-implemented inventions. In 2012, *Mayo Collaborative Services v. Prometheus Laboratories*[89] limited certain methods used for medical diagnosis as inventions, disqualifying them from patents as a "law of nature."

In 2013, the court handed down *Association for Molecular Pathology v. Myriad Genetics*,[90] ruling an isolated gene, in this case one highly correlated with breast cancer's presence, is a "product of nature." The court disregarded human intervention with the gene, used as a diagnostic tool. This decision further narrowed patent eligibility. In 2014, *Alice Corp. v. CLS Bank*[91] deemed a sophisticated financial software-implemented invention to be ineligible for a patent, labeling it an "abstract idea."

With its resulting "*Alice-Mayo* Framework," the Supreme Court "effectively rewrote the statute," former Federal Circuit Chief Judge Paul Michel and IP attorney Matthew Dowd write. This legislating from the bench "appears to be unsupported by precedent" and relies "on unsupported and incorrect characterizations about patents and the [c]ourt's own precedent."[92]

The *Alice-Mayo* Framework has thrown patent eligibility jurisprudence into confusion. The court has invoked undefined terms, such as "abstract," and provided clunky guidance for determining eligible subject matter. As a result, courts have little to go on in patent eligibility determinations, while every estimation in case after case is open to interpretation and contradictory reinterpretation. Instability and uncertainty result, leaving section 101 cases in the depths of conflicting rulings. Former PTO Director David Kappos told the U.S. Senate, "The Supreme Court, Federal Circuit, district courts, and USPTO are all spinning their wheels on [patent eligibility] decisions that are irreconcilable, incoherent, and against our national interest."[93]

Together, these and other rulings encroach on the plain language of the law and close off what formerly was and should be regarded as patent eligible subject matter—i.e., intellectual property. Madigan and Mossoff report the damage *Alice-Mayo* has done. From 2014 to 2017, 60 percent of patent eligibility rulings in federal district court or before the Federal Circuit invalidated the patents (PTAB invalidations were even higher).[94] Moreover, of nearly 18,000 patent applications PTO rejected finally, the European Patent Office, China or both granted patents on about 10 percent of the same or similar inventions. Madigan and Mossoff cite several examples of rejected applications, one a diagnostic method for liver disease invented at the University of California; both China and Europe granted the patent.[95] This diminution of threshold eligibility under section 101's broad language undermines the statutory presumption of validity adjudicators are supposed to extend to patented inventions. Thus, unsettled patent eligible subject matter contributes to the gold standard American patent system's fading as other nations' patent systems comparatively gain.

Injunction and Efficient Infringement

The essence of patent rights is the right to exclude. An injunction makes that right meaningful. Fundamentally, patents secure the right to keep others from making, using, selling, or importing an invention without the permission of the rights holder.

Before 2006, federal courts that found a patent valid and infringed routinely issued an injunction. That was because, typically, monetary damages alone won't fulfill justice in property rights matters. Getting a squatter removed from one's home is meaningful relief, not just a court ordering the squatter to pay rent for the period of past unlawful residence and for future residence. Therefore, in patent cases, courts usually granted injunctive relief.[96] Courts permanently enjoined proven infringers from continuing commercial activity, such as making and selling knockoffs of the patent owner's invention. Injunction is in addition to money awards.[97]

After two centuries of permanent injunctions in 95 percent of patent infringement cases, the U.S. Supreme Court's 2006 *eBay v. MercExchange*[98] effectively led to courts using a categorical rule denying injunctions. Patent owners now get little help from courts in exercising their right to exclude by injunction.

What caused this 180-degree reversal of pro-property owner justice, leaving a nearly impossible categorical rule denying injunctions? Justice Clarence Thomas's *eBay* opinion was unanimous and not disruptive; Chief Justice John Roberts's concurring opinion affirmed the historical trend. Justice Anthony Kennedy's opinion urged courts to disregard established patent rights due to alleged "patent trolls," overstated patent litigation, and "weak patents." Kennedy suggested making distinctions among patent owners. The resulting categorical rule against granting patent injunctions hit licensors especially hard.[99]

Thus, *eBay*, as courts have come to apply it, weakens patent rights by denying the central property right, the right to exclude. This leaves patent owners without the ability to force infringers to negotiate, negotiate in good faith, and agree to terms. Without injunctions, infringers gain an unfair advantage. They can freely keep making, selling, using, and importing knockoffs. Infringers continue collecting ill-gotten gains from this infringing activity. They continue to cut into the patent owner's market share and profits, with no sense of urgency to do anything other than keep bleeding the innovation's rightful owner. This devalues patents, "a zero value for exclusivity in a license deal," which Osenga shows has led to many fewer exclusive licenses.[100] The growth of nonexclusive licensing at exclusive licenses' expense is a canary in the coal mine of patent rights.

"The *eBay* decision has substituted a compulsory licensing scheme for an exclusive right. [Absent a permanent injunction,] the only thing to discuss with infringers is the price of a de facto compulsory license, because the property owner can no longer boot the trespasser from his property."[101]

Predatory infringement by Big Tech firms plagues, among other innovative, IP-intensive companies, Netlist, a small, innovative firm in the advanced semiconductor space that holds more than a hundred patents. Netlist's chief executive Chuck Hong tells about the struggle to defend Netlist's patents against infringement, writing in *Fortune* an essay headlined, "Big Tech's Abuse of the Patent System Must End—Take It from Me, I've Fought Google over IP for Years." Netlist has faced knockoff technologies, multiple years-long challenges of its patents, and being ganged up on by allies of the infringer Google.[102]

* * *

If you're getting the idea that the drip-drip-drip weakening of America's patent rights over the past several decades has empowered patent infringers while denying inventors and patent owners of biblical ownership rights, you're correct. The new situation is asymmetric, heavily favoring companies with deep pockets and willful disrespect for anyone else's intellectual property.

The situation is such that *The Economist* reports, "Boris Teksler, Apple's former patent chief, observes 'efficient infringement', where the benefits outweigh the legal costs of defending against a suit, could almost be viewed as a 'fiduciary responsibility,' at least for cash-rich firms that can afford to litigate without end."[103]

"Efficient infringement," also known as "predatory infringement," amounts to established firms, whose devices or platforms could incorporate or market an innovative smaller firm's invention under a license, feign interest, take the IP, then lose interest in striking a deal as they trot out a copycat of the invention under their own names. The innovator's invention could have enhanced and added value to the infringer's product or offerings, while giving the smaller firm access to developed markets and scaled-up sales. The established firm would have sold more of their own products with the licensed IP—a win-win. Predatory infringement deprives the innovator of licensing revenues to repay investors, pay workers, cover operating costs, and fund R&D.

Predatory infringement leads to endless litigation in multiple forums, such as federal court, PTAB, and the ITC. For instance, cybersecurity entrepreneur Centripetal Networks found itself in this situation with Cisco and its allies. Centripetal has litigated infringement in district court, the Federal Circuit, and repeatedly defended its patents' validity at PTAB.[104] Sonos, which invented wireless speakers, has pursued Google and Amazon for patent infringement in multiple litigations in such forums.[105] Medical device firm Masimo has waged legal battles in several adjudicatory forums against Apple for patent infringement of Masimo's wireless blood-oxygen monitor. A knockoff was added to the Apple Watch without a license.[106]

Centripetal prevailed in district court. The court awarded multibillion-dollar damages because of "an egregious case of [Cisco's] willful misconduct beyond typical infringement."[107] That was far from the end of litigation, which has continued in several forums. The appeals court tossed the judgment on a technicality. Sonos has continued its litigation on several fronts, including at the ITC and Federal Circuit.[108] Masimo prevailed against Apple's infringing watch models, with the ITC issuing an import ban that the Federal Circuit upheld.[109] Other Masimo litigation has continued, with Apple unlikely to take a license.[110]

• • •

Able to bring PTAB and levels of federal court to bear in repeated challenges against patent owners, David versus Goliath contests exist. Only, in the predatory infringement version, Goliath usually wins. Asymmetric lawfare plays out in this near-predetermined way because the AIA handed predatory infringers advantages: easily invalidated patents in PTAB proceedings, inordinately expansive prior art, and other loopholes. Courts add wholesale instability from the *Alice-Mayo* Framework and, most damaging to those enforcing patents, judicial bias against granting permanent injunctions, even though patent validity and infringement have been proven.

Oil States and IP Rights

Since the AIA's enactment and its creation of PTAB, courts—including the Federal Circuit and the U.S. Supreme Court—have rationalized the administrative branch's quasi-judicial body for canceling patent rights. They claim these consistent with the Constitution and its biblical, Lockean property rights the Founding Fathers intended and enacted in the first federal patent statutes. An extensive record of judicial rulings in patent cases developed over a two-century period, upholding patents as private property rights.[111] In 2018, the Supreme Court ignored all that.

In *Oil States Energy Services v. Greene's Energy Group*,[112] the court asserted PTAB's inter partes review proceedings are constitutional under "public rights doctrine." The 7–2 majority opinion, penned by Justice Thomas, called patents a "public franchise," just like any other benefit the government confers and may withdraw. *IPWatchdog*'s Quinn responded to the ruling the very same day. He wrote, "If patents are public rights that can be challenged at any time and revoked, that necessarily means they are not property."[113]

"It is no exaggeration to remark that the *Oil States* Court ignored a substantial amount of precedent that clearly described a patent as a private property right," Michel and Dowd say. "Thus, for decades, the Supreme Court seemingly confirmed the view that a patent secures a private property right—a right based on the fruits of one's labor and granted as part of the quid pro quo for disclosing the invention to the public. *Oil States* squarely rejected that view" (internal citations omitted).[114]

The high court relegated patents to invalidation determinations made by an administrative tribunal, rather than reserved for federal court, as are matters regarding other forms of property. *Oil States* gives the Supreme Court's imprimatur to downgrading secure ownership rights to one's invention—a central characteristic of U.S. patenting 1.0, 2.0, and 3.0.

Oil States has countenanced the administrative power to regulate and even withdraw the inherent right to own what one creates, to control and benefit from the fruits of one's labor.

• • •

Essentially, the *Oil States* ruling represents the undoing of America's commitment to the biblical creativity-ownership model. The other ways in which patents and IP rights have been undermined—PTAB, imbalanced inter partes review, unsettled patent subject matter eligibility, refusing injunctive relief against proven patent infringers, weaponization of antitrust against IP rights—flow from postmodern distortions and eradication of what made American invention, secured by reliable property rights, unique and powerfully effective. The fact that this wholesale abandonment of private property rights in one's inventions comes from a "conservative" Supreme Court underscores the sea change.

Borrowing Madigan and Mossoff's apt metaphor, the American patent system has nearly devolved from the biblically grounded golden era of U.S. patenting 3.0 to the leaden era of U.S. patenting 4.0. Moreover, Bork's observations of troubling trends in antitrust—movement away from democratic political decision making to courts' lawmaking, from free markets to regulated markets, from general welfare to group welfare, from liberty and merit-based reward to equality of outcome and reward based on status, i.e., "equity"—also mark U.S. patenting 4.0.

President Biden's aggressive antitrust agenda,[115] especially its several aspects aimed at reducing IP rights,[116] and the fact it employs bureaus of the Administrative State to accomplish strong-antitrust, weak-patents goals only furthers the patenting 4.0 model's taking root. The United States pursued this path in the mid-twentieth century. Yet, we have forgotten those policies' adverse effects. They distorted market forces and diminished dynamic competition; fostered large-firm dominance and impeded nimble, small innovators; and demonstrated the flaws and failings of central planning.

If we fail to remember the past, then in Santyana's words, we "are condemned to repeat it."[117] In this case, it means the greatest patent system in the world has been reduced to a shadow of itself while America's competitors in China, Europe, and elsewhere proceed toward passing the

United States as global leaders in cutting-edge, high-value invention and innovation.

We are sliding from U.S. patenting as virtuous circle to vicious cycle. Judge Michel's sobering assessment is "the U.S. patent system was routinely referred to as the 'gold standard.' But those days are gone" (internal citations omitted).[118] America will have squandered God's Common Grace blessing for human flourishing. We soon may repeat for our once-great patent system David's lament over the death of his close friend Jonathan and King Saul, "How the mighty have fallen!" (2 Samuel 1:19b).

Chapter 17

Restoring the Biblically Based Creativity-Ownership Link

*"Progress means getting nearer to the place you want to be.
And if you have taken a wrong turning, then to go forward does not
get you any nearer. If you are on the wrong road, progress means doing
an about-turn and walking back to the right road; and in that case
the man who turns back soonest is the most progressive man."*
— C. S. Lewis[1]

*"There are no wealthy countries with weak patent rights,
and there are no poor countries with strong patent rights."*
— Stephen Haber[2]

For those who understand how human law and public policy can embody and apply divine communicable attributes—such as creativity and ownership—the Creator of the universe has endowed in human beings, the task at hand is to reclaim that divine birthright. Human creativity and ownership, together, are the essence of the formula for human flourishing. Creativity and ownership are necessary components for individual human beings and humanity to thrive.

Yet, the United States has allowed these two vital elements to become severed in U.S. intellectual property—especially patent—law and policy. The solution: Reestablish the link between creativity and ownership. If what this bond yields is to be stable for the long term, this link must firmly reset the biblical underpinnings of the Constitution's Article I, Section 8,

Clause 8 and the core components of the body of patent laws that existed at the time of America's "golden age of patenting" from the 1830s until the 1930s and reprised in the 1980s.

We face a restoration job similar to Nehemiah's rebuilding Jerusalem's decrepit walls. His task was about much more than simply renovation and construction. Changing the people's hearts and minds had to happen too. And so it must today.

Here, I want to clarify something. Below, I make certain policy and legislative recommendations. Such policy prescriptions are prudential judgments, similar to different Christian denominations or churches adopting different methods of baptism or administering the Lord's Supper. I have sought the whole counsel of Scripture in reaching my recommendations. Yet other permissible options may exist that also accord with biblical principles of creativity and ownership. Only the Lord is omniscient. I do believe these recommended here are the best policy options for reconnecting secure property rights in one's creative or inventive works, based on Scripture, empirical and historical evidence, and reason.

In order to restore and cement the link between biblically based creativity and ownership, three layers need attention. Each level has suffered degradation and disrepair. These layers are foundational tenets, the innovation ecosystem, and IP law and public policy.

Foundational Tenets

In 1987, President Reagan released an Economic Bill of Rights. It was comprised of four freedoms: the freedom to work, to enjoy the fruits of one's labor, to own and control one's property, and to participate in a free market.[3] These freedoms and their corollaries make up the layer of foundational tenets in need of restoration today. They relate to creativity and ownership but extend beyond those qualities.

Reagan was referring to four pillars of the social contract. They derive from the Western canon, including Judeo-Christian Scripture.[4] We have seen these weight-bearing columns of human society throughout our examination of biblical creativity and ownership, such as in chapter 12.

Freedoms associated with economic endeavors, as Reagan parses them, include a free enterprise economic system. The right to work ("the right to contract freely for goods and services and to achieve one's full potential without government limits on opportunity, economic independence, and growth") and the right to enjoy the fruits of one's own labor ("the right to keep what you earn, free from excessive government taxing, spending, and borrowing") are facets of free enterprise and private property. Ownership ("the right to keep and use your property, free from government control through coercive or confiscatory regulation") is property rights.

Rightly construed, property rights are broad. The right to private ownership of property applies to property of many species—land, buildings, one's earnings and profits, financial instruments, one's inventions and creative works, and personal possessions, for instance. Freedom beyond economic liberty means individual liberty to govern oneself, freedom to take creative initiative or apply ingenuity, exercise free speech, free association, and free religious belief, for example.

Corollaries of these pillars are the God-ordained human institutions of just society, such as the smallest society, the family; community; and state. Each of these levels of society stands accountable to God. Romans 13:4 (NIV) calls civil authorities "God's servants, agents of wrath to bring punishment on the wrongdoer." From this divine provision follow Common Grace provisions, such as ordered liberty, the rule of law, equal justice under law, consent of the governed, fairness, and due process, which provide boundaries to constrain evil. These lay the foundation for healthy society. Moreover, private property rights underlie ordered liberty, the rule of law, and the rest.

Bolstering the integrity of diminished elements of the United States' constitutional order, making these cornerstones secure, is paramount. Thankfully, much of the U.S. constitutional structure remains relatively intact. However, where it has been distorted or weakened, restoration is necessary. Some examples of what needs addressing: unaccountable bureaucracy—the Administrative State of which the Federal Trade Commission, PTAB, and the Consumer Financial Protection Bureau are prime examples—; abuse of presidential executive orders as broad, policy-changing vehicles; and courts' legislating whenever judges (both judicial branch appointees and administrative functionaries) exceed their designated powers. Instances of these arrogations have occurred with respect to patents and patent rights.

To safeguard the foundational tenets of civil government in our constitutional republic, the legislature should return to the fore as the sole branch of government with lawmaking powers, and its checks on the excesses of the executive and the judiciary should be strengthened. This is mainly a matter of restoring legislative muscle memory, exercising will, and strengthening backbone to stand against excesses of the other branches of government. Personnel is policy, thus carefully vetting devotion to the fundamental principles of the American social contract must be a top priority as voters scrutinize candidates. Also, executive and legislative officials must examine potential appointees to government positions. Legislative tools of oversight, the purse strings of the federal treasury, and removal or other sanction for exceeding one's legitimate authority should be employed when necessary. These measures factor into preserving and protecting the foundational tenets of our republic. Such faithfulness to one's duty of office and vigilance toward officials' conduct will strengthen citizens' property rights generally and the creativity-ownership combination in particular.

Innovation Ecosystem

The second layer to address for reconnecting ownership with creativity is the U.S. innovation ecosystem. This infrastructure is comprised of components that foster and facilitate invention and discovery. Among them are access to venture capital (VC), resources like suitable facilities for pursuing innovation, and qualified workers.[5] Other innovation resources include industrial, commercial, and academic entities engaging in complementary sectors. Some perform basic research, others applied research, others manage intellectual assets and transfer technology, and others provide supplies and services to innovators and commercializers.

Start-up incubators and business accelerators offer shared workspace, laboratory facilities, mentorships, and networking opportunities as part of a locality's innovation ecosystem. Fabrication and engineering firms capable of producing prototypes of inventions, law firms skilled in business and IP law, and technical schools offering apprenticeships and cooperative training opportunities belong to local or regional innovation ecosystems. Technical colleges have become an especially important source of qualified technologists for regional industry. Students come equipped with diplomas and certifications, oftentimes with hands-on experience in industrial settings. Economic development agencies, nonprofits such as inventors' clubs and associations focused on local industrial sectors, and contract labs and manufacturers help round out innovation ecosystems. Some may sponsor networking events and job fairs, hold colloquiums and workshops, or host contests where entrepreneurs pitch their start-ups and the innovations they seek to commercialize. Winning contestants usually get cash prizes.

The proof of the importance of an innovation ecosystem appears, for example, in Boston around the Massachusetts Institute of Technology (MIT); Silicon Valley, with its start-up track record and collaboration with universities such as Stanford; Reston, Virginia's high-tech firms; and Maryland's life sciences corridor from Baltimore's Johns Hopkins to federal biomedical facilities in Metro Washington. Congressman Thomas Massie, an MIT alumnus and tech transfer success story, has noted the total revenue of MIT's start-ups constitute the thirteenth largest economy (by gross domestic product) in the world.[6]

Innovation ecosystems proliferate throughout the United States. Some thrive and box above their weight. Many local or regional innovation ecosystems are more modest by comparison to the MIT and Silicon Valley innovation ecosystems but are impactful locally. They all are necessary, if not sufficient.

• • •

Among the vulnerabilities threatening U.S. innovation ecosystems across the board is falling access to capital investment. The Alliance of U.S. Start-ups & Inventors for Jobs (USIJ) documents how weakened IP and the instability of the U.S. patent system cause venture capitalists to retreat from investing in patent-centric technologies. USIJ reports, from 2004 to 2017, VC funding plummeted from more than 50 percent going into patent-intensive startups, such as those in biotech, semiconductors, and pharmaceuticals, to 28 percent. Patent-intensive enterprises received only 30 percent of VC funds between 2013 and 2017. In that period, 70 percent went to non-IP-dependent startups, such as those developing social network-platform, consumer finance, hospitality, and leisure offerings.[7]

Capital formation resources are a vital element of an innovation ecosystem but they are diminishing. This harms those pursuing sophisticated innovation and the ecosystem. Investors understand the heightened risk of losing rather than profiting from their financial investments that increasingly unreliable, unenforceable patents pose. Thus, VCs have reduced their investments' risk exposure.

This rational decision expands the consequences of the "valley of death," the time between launching a commercialization enterprise and its failure, bankruptcy, or market success. USIJ's VC investment report documents how weak IP destroys access to capital for those with more significant inventions, while start-ups with less impactful, even frivolous products or platforms secure investments. Fewer life sciences, advanced manufacturing, computer hardware, and microchip inventions will be developed, come to market, save or improve human lives and health, and have the opportunity to create new jobs, new markets, and dynamic competition. This because of unreliable, unenforceable U.S. patents.

Rep. Massie explains how secure property rights in IP relate to investment. He says "it takes machinery and equipment to produce inventions, and people will not make the investments in that machinery to duplicate the invention if they know that the idea can be stolen from them and they'll be left with a bunch of equipment they can't use that they've invested in."[8] Hence VCs' de-risking their investments away from patent-oriented enterprises is rational. It rests on the fact that property rights in these areas have become tentative; the foundational tenets that should buttress patent property rights no longer are assured in the United States.

With less investment funding available to patent-centered start-ups, innovation ecosystems may not form, flourish, or realize key parts of their innovative potential. Attracting VCs and other investors as prominent players in the innovation ecosystem should be a priority. To coax investors back to funding IP-intensive enterprises, particularly patent-centered start-ups, requires restoring the durability, robustness, and secure private property rights of patents. Reestablishing the biblical and Common Grace link between creativity and ownership for inventors and patent owners will help bring certainty and extend the virtues of that link to investors and others who round out the constellation of the innovation ecosystem.

Patent Law and Policy Reforms

The third layer, which the previous chapter established is in desperate need of realigning with the biblical creativity-ownership model, is fixing U.S. patent law and public policy. As we've seen, Congress, the administrative branch, and the judiciary have each made patent policy worse for innovators and favored patent infringers. Congress has enacted patent-weakening laws like the America Invents Act (AIA). Executive agencies have swung like a pendulum between pro-patent and anti-patent policies. Courts have wreaked havoc in core areas of patent jurisprudence over recent decades.

Executive agency patent policies have swayed between agency capture by weak-patent advocates leading the PTO, the National Institute of Standards and Technology (NIST), the DOJ Antitrust Division, the U.S. Trade Representative (USTR), and the Federal Trade Commission (FTC) in one administration, and political appointees with pro-innovation policy views leading the next administration's agencies. One administration puts in place rules and guidance that a change of administration reverses with oppositely directed rules and guidance. This situation is untenable. Cyclical executive action bringing temporary, 180-degree policy changes greatly disrupts the efforts of those engaged in the U.S. innovation ecosystem, whose R&D and commercialization schedules don't align with the election calendar.

The best course is to stop the merry-go-round of dueling patent policies. The primary cure rests with Congress. It is the branch most directly accountable to the American people. Because patent legislation and policy do not divide along party lines, the legislative branch stands in the strongest position to fix what ails the U.S. patent system. Furthermore, Congress can (and should) prescribe the specific policies that executive agencies carry out regardless of which party currently holds the White House, and can constrain courts through legislation. Moreover, placing specific patent policies in black-letter law reduces the likelihood back-and-forth policy by fleeting executive actions continues. A statute stands after whichever administration presently holds the reins of executive government. Also, courts will apply those laws, building a body of case law whose precedents perhaps will align with the policy goals of legislators who crafted the statute.

Thus, the legislative branch is the key to the restoration of biblical and historical American patent rights, of moving us from U.S. patenting 4.0 to 5.0.[9] This course won't be easy; passing an act of Congress requires dedication, coalition-building, and heavy lifting. Meanwhile, Congress holds powers by which the legislature can tug the reins of the other two branches. Annual appropriations of agency and judicial funding, riders restricting agencies from expending money on certain programs or policies; oversight and investigatory tools, the powers of subpoena and contempt of Congress charges, removing courts' jurisdiction over certain subjects, and the ability to remove officials give Congress leverage to compel compliance by bureaucracies and courts.

* * *

What laws and policies should Congress change to bring about U.S. patenting 5.0, a revival of the "golden age of patenting" for the twenty-first century? Several legislative priorities are in or related to patent law, with other changes needed in antitrust policy.

Patent Legislation

The foremost legislative remedy for strengthening patented inventions' value, patents themselves, and patent exclusivity would be to enact legislation restoring a ***presumption in favor of permanent injunction*** in cases where courts have found patents valid and infringed. This would overturn the destructive *eBay v. MercExchange* ruling.

Legislation such as the Restoring America's Leadership in Innovation Act (RALIA) and the Realizing Engineering, Science, and Technology Opportunities by Restoring Exclusive (RESTORE) Patent Rights Act would require courts that have made such determinations to presume continued infringement will cause irreparable harm or that injunction is in order. This would force the proven infringer to halt commercial activities involving the infringed invention. Injunctive relief would not affect monetary damages the court awards.

"One of the first things we should do is restore the ability to get an injunction if somebody is stealing your [patented invention]," Rep. Massie says. "The current state of affairs is that even when they're caught, [infringers] get to keep stealing it."[10]

The restoration of a high likelihood of infringed patentees obtaining injunctive relief would shift the dynamics. Property owners would have a level negotiating table with economically constrained, illicit users of their inventions. Infringers could no longer hold out on negotiating a fair license. They would be motivated to strike an agreement as soon as possible because the injunction has halted their revenue stream of ill-gotten gains. "Generally speaking, the greater the threat of an injunction, the greater the bargaining leverage enjoyed by the patent owner in negotiating license terms, and vice versa," write Barnett and Kappos.[11] These experts "encourage Congress" to legislate the "restoration of the historical presumption favoring injunctive relief for prevailing patentees, which would more directly correct the underdeterrence effect."[12]

• • •

Congress should amend the Patent Act's section 101 to restore broad ***subject matter eligibility*** for patents. "We shouldn't say that patents only apply to some fields of invention," Rep. Massie, the inventor, patent owner, and entrepreneur, says. "Technology by definition changes. . . . Technology is new things. And there's this tendency recently to say, 'Well, computer software is a new way of inventing and we shouldn't allow patents in this domain because you can invent so quickly in this domain by merely changing the code. You don't need to develop tooling.' Well, I think that's bad because . . . you're going to disincentivize invention in other fields . . ."

Mossoff explains the flawed reasoning Rep. Massie criticizes. "The value in a software program is the functionality of the program. . . . The functionality of binary code in a specific computer program is in principle no different from the functionality achieved in the binary logic hardwired into computer hardware. . . . [It's] the digital equivalent of 'a specific machine.'"[13]

Legislation to cast aside the errant rulings (e.g., *Bilski*, *Mayo*, *Myriad*, *Alice*) and their progeny should eliminate all judicially created exceptions to patent eligibility, forbid courts from creating future exceptions to what is patent-eligible, and prohibit courts from considering substantive patent considerations—an invention's novelty, usefulness, or obviousness—in connection with the threshold question of subject-matter eligibility.

To reestablish the "anything under the sun that is made by man" breadth of patent-eligible subject matter and reverse all that "has changed dramatically over the last fifteen years,"[14] we should base section 101 determinations on a claimed invention as a whole. Assessments of individual patent claims should be reserved to patent examination on the merits, not the initial eligibility inquiry.

Rep. Massie's RALIA and Sen. Thom Tillis's Patent Eligibility Restoration Act each would eliminate the *Alice-Mayo* Framework, end courts' narrowing of patent eligibility, and provide clear congressional intent.

• • •

As discussed in the previous chapter, the Patent Trial and Appeal Board (PTAB) invalidates patent claims using procedures, rules, and processes that tilt the outcome as all but determined to cancel patents. We contrasted PTAB with the procedural fairness and due process that federal courts and the U.S. International Trade Commission (ITC) observe. The America Invents Act's "creation of trial-like PTAB proceedings is premised on the efficiency of substituting administrative process for judicial process. Yet a wasteful glut of duplicative litigation persists because of gaps and misalignments in how the PTAB is structured relative to the district courts."[15]

Several approaches have been proposed for dealing with PTAB. One is to **abolish PTAB**. This would keep all adversarial patent validity disputes in federal courts, while restoring the option of de novo review by federal courts and juries. That leaves non-adversarial patent reexamination at PTO. RALIA takes the course of repealing PTAB. RALIA sponsor Rep. Massie says, ". . . I think our Congress made a big mistake when they created the PTAB, . . . a court that is not in the judicial branch. And the result has been horrific for inventors and for the assumed validity of patents." PTAB repeal would "get [the United States] back to having patents adjudicated in courts instead of . . . the executive branch."

Another approach, found in Sen. Chris Coons's Promoting and Respecting Economically Vital American Innovation Leadership, or PREVAIL, Act, would preserve PTAB but impose reforms causing PTAB proceedings to operate much like Article III courts. **PTAB procedural and due process reforms** include requiring standing (i.e., the one bringing a challenge before PTAB must be directly affected by the patent at issue); meeting the higher standard of proof, clear and convincing evidence that courts use in property cases; codifying the same standard for interpreting patent claims federal courts use; tying various parties to the AIA's one-year time bar, which has become a loophole for multiple patent challenges against the same patent; and requiring PTO's quasi-judicial administrative proceedings to defer to the judicial branch when a court determines a patent is valid.

Another approach would ***require PTAB to get the consent of a patent owner*** whose patent is exposed to validity challenge in PTAB. Withholding consent to PTAB proceedings would put the validity determination in federal court. This approach appears in the Balancing Incentives Act, sponsored by Rep. Marcy Kaptur.

Another idea is to require PTAB to pay patent owners, whose patent claims the panel invalidates, ***just compensation***. Compensation monies should be drawn from PTO's funds. Every PTAB panel that decides to cancel a patent, entirely or in part, should have to determine the amount of compensation that's just, based on real-world, market-based data, R&D investment, and time remaining in the patent term. The Fifth Amendment guarantees just compensation whenever the government takes private property, which is what PTAB does. By making PTAB compensate harmed patent owners and requiring it to bear the costs of property-linked compensation, the government agency that takes private property would incur the costs of its decisions.

Dismantling PTAB and its quasi-judicial proceedings aligns more closely with divine creativity-ownership principles as well as original American patent principles, practices, and jurisprudence. This is the best approach from the perspective of its merit in principle. The challenge would be overcoming the many vested interests and political apologists dedicated to protecting PTAB.

Mending instead of ending PTAB by placing court-like fairness and due process, requiring patent owner consent to proceed, or enforcing just compensation for invalidations would also be a stiff challenge. Given that AIA advocates sold PTAB as an alternative forum for litigating patent validity, the choice was supposed to be either federal court *or* PTAB, *not* both court *and* PTAB.[16]

Requiring patent challengers to choose one forum or the other, with the backstops of reformed PTAB policies, procedures, and standards mirroring those of federal courts and the ITC; consent; just compensation; or a combination, would give patent owners meaningful relief from PTAB's current gamesmanship and inherent unfairness. Such reforms may diminish PTAB's usefulness to its users to such a degree inter partes review petitions fall. That outcome could ease the prospect of eliminating PTAB or at least its adversarial proceedings, as they become redundant of judicial litigation and PTAB amasses comparable invalidation statistics.

These approaches to dealing with PTAB as it exists today represent progress. Each one squares with C. S. Lewis's observation, "progress means doing an about-turn and walking back to the right road."[17] The right road and right direction here is the biblical model of private property rights the Founders adopted in the American patent system.

• • •

Restoring robust private property rights in one's inventions, creations, and IP is imperative, in light of the Supreme Court's *Oil States* decision. Quinn calls *Oil States* a "stupid ruling" because in it the high court "acknowledged what [it] had done to the patent system. [It] acknowledged what Congress had done to the patent system. They turned the patent right from a property right into a government franchise."[18]

To overrule *Oil States*, the patent statute should state that **inventions and patents are private property and patent rights are private property rights**, on par with other property. This includes patent owners' right to de novo review in federal court. The revised language should affirm patents as securing exclusive rights, including the right to exclude and the rights to assign, license, or dispose of the patent. RALIA contains such a provision.

• • •

After more than forty years of successfully leveraging secure patent rights, clear title to inventions, and democratized, decentralized decision-making about technology management, the Biden-Harris administration in 2023 set in motion an initiative to dramatically change the Bayh-Dole Act model. NIST proposed a framework for considering the price of a product in a federal agency's decision whether to exercise march-in rights on a patent. Any patent connected at any point with federal funding becomes subject to expropriation based on how the government regards the "reasonableness" of the market-set price.[19]

This creates tremendous uncertainty over how secure patent rights are if federal funding underwrote a product's initial research. Former PTO Director David Kappos warns of the response to this move: "What I'm going to be telling companies that I advise, and I expect every other lawyer in the nation is, stay away from federally funded R&D. It's now toxic. It's contaminated. Don't touch it."[20]

This action was allegedly about reducing the price of prescription medicine.[21] However, it quickly became clear very few pharmaceuticals would be exposed to march-in, which is where a university's or a small business's license could be seized and given to another firm, including a competitor. Most drugs are protected by several patents, most of which cover inventions developed with private investment after commercialization began.

The policy began coloring companies' interest in public-private partnerships, federal research funds, or otherwise exposing themselves to the prospect of investing private resources in commercializing IP and bringing to market a product whose IP the federal government could take away in the future.

Before this, Bayh-Dole had proven itself to be well worth the return on the comparatively meager federal "investment." The practical uses of inventions derived from those basic research grants more than repay taxpayers. However, in 2001, a false premise asserted that Bayh-Dole march-in could take into account the price of a patent's commercialized product. This fiction aimed for march-in use as a government price control on prescription drugs. Senators Bayh and Dole vigorously denied the false assertion. Bipartisan administrations have consistently dismissed petitions demanding march-in—Biden's National Institutes of Health (NIH) in March 2023 rejected a march-in request against the drug Xtandi, shortly before the administration's "reasonable pricing" initiative.[22]

The president ultimately ***withdrew the 2024 "reasonable pricing" framework,*** after losing the nomination to Vice President Kamala Harris, and candidate Harris lost the presidential election. Eliminating the NIST Bayh-Dole march-in framework should help restore the law's features and the accurate interpretation of the statutory grounds for march-in that brought about Bayh-Dole's successful forty-plus-year tech transfer record. However, with ten days to go in the Biden administration, the National Institutes of Heath issued guidelines for licensing NIH patents, doubling down on product pricing and other noncommercial requirements. These NIH guidelines will chill NIH licensing—just as NIH's 1990s failed experiment of "reasonable pricing" contractual terms repelled would-be licensees who would have liked to attempt to commercialize a given invention.[23] Calling George Santyana!

* * *

Several other patent reforms would secure the U.S. patent system's biblical creativity-ownership moorings. Each would return specific aspects of patent policy and practice that distinguished the American system for most of its existence.

We should **replace first-inventor-to-file with the first-to-invent** criterion for determining which competing inventor is granted a patent.[24] The first person who invents something is its rightful owner, as Chief Justice Marshall's "inchoate and indefeasible property . . . vested by the discovery" describes, and should receive priority based on the earlier date. This restores the creativity-ownership rule the Founders established.

We should **stop publishing patent applications before patent grant**. The American Patent Bargain is sharing an invention's details in exchange for exclusivity. Historically, that began only when the patent was granted. Automatic patent publication before issuance distorts the bargain. It cuts into exclusive ownership, destroying exclusivity if patent grant doesn't occur. If PTO publishes a patent application eighteen months after its filing and the application is denied, the denied application becomes prior art. This robs the inventor of other options such as trade secrecy.

Pre-issuance publishing hands China and other foreign rivals a head start on commercializing and developing a market for the new invention. In addition to realigning the U.S. patent system with divine invention-ownership principles, restoring confidentiality at PTO would serve U.S. national security, economic, and competitiveness interests.

We should **eliminate prior user rights**. This loophole benefits patent infringers. Ending it could incentivize due diligence through "freedom to operate" patent searches to avoid infringing someone's patent. Its demise may increase seeking licenses to others' patents, undermining the practice of "efficient infringement."

Another creativity-ownership reform would be to **roll back the AIA's expansive redefinition of prior art** and **limit prior art considered in U.S. patent examination. Only** U.S. patents and English translations of foreign patents and applications easily accessible from the United States should be accepted as prior art in U.S. patent examination and PTAB or other U.S. patent proceedings. Pre-AIA prior art standards for obtaining a patent make more sense. U.S. patents can only be enforced in the United States, so it's unreasonable to exceed U.S. boundaries for obscure foreign prior art beyond what is easily discoverable by an American.

This limitation should accommodate a **one-year grace period** as the law did in the United States before 2011. Restoring pre-AIA prior art rules would guard U.S. patentees against exposure to dubious or fraudulent prior art claims. Also, we should **restore the one-year grace period** from first public disclosure about one's invention until a patent application must be filed.

We should **make the U.S. patent term twenty years from the date a patent issues**. This change would restore the historical U.S. policy of patent terms beginning upon grant. It would also spare American inventors the loss of patent term ticking away during the often-lengthy examination period.[25] By exceeding the TRIPS Agreement minimum of twenty years from filing, this directly benefits American inventors and sets an example for other TRIPS nations.

The United States should initiate **repeal of the 2022 TRIPS waiver**, the policy that threatened to expropriate COVID-19 vaccine IP. The World Trade Organization in 2024 stalemated over a proposed expansion of the TRIPS waiver, which would expose IP covering COVID diagnostics and therapeutics.[26] A U.S. and allied effort should insist on raising WTO's criteria for any future consideration of waiving TRIPS to a high **evidence-based standard**. It should be akin to clear and convincing evidence, with a presumption against waiving IP rights. Any country seeking to waive TRIPS should bear the burden of proof and obtain a high supermajority.

We should **safeguard the U.S. International Trade Commission**, whose IP-related trade procedures and rules are impartial and evidence-based. The even-handed, rules-based trade panel has become a target of deep-pocketed importers of patent-infringing, foreign-made goods, from Big Tech firms to automobile makers. The ITC must remain a bulwark of fairness and due process with the power to block importation of products infringing U.S. innovators' patents.

The Restoring America's Leadership in Innovation Act would address many of these pro-creativity-ownership issues. RALIA would restore access to injunctions against patent infringers, repair patent-eligible subject matter breadth, repeal PTAB, restore U.S. patents' private property status, resurrect the first-to-invent standard and the one-year grace period, and halt PTO from publishing U.S. patent applications unless a patent has been granted.

Antitrust Policy

We need appropriate balance at the intersection of intellectual property and antitrust. This would help reduce the radical swings of the politicized pendulum phenomenon with each change of administration and guard against property rights being made subservient to antitrust.

First, Congress should **write into law the New Madison Approach**, in which the dynamic competition resulting from the commercialization of patented innovation tempers antitrust enforcement. This policy would require antitrust enforcers to stop treating the exercise of exclusive patent rights as if it were a per se antitrust violation. Placing New Madison in black-letter law would give it greater permanency. It would require competition agencies to apply this approach in cases where patents are involved in alleged anti-competitive conduct. It also should prevent a repeat of misguided cases like the FTC's 2017 lawsuit against Qualcomm that the FTC lost on appeal nearly four years later.

President Reagan's Young Commission captured the rationale for what later was dubbed the New Madison Approach: "The very act of [IP] licensing is procompetitive rather than anticompetitive . . . [and justifies] view[ing antitrust] restrictions in light of all the surrounding circumstances, especially the impact on competitiveness. . . . Not only do licenses introduce more competitors into the marketplace, but insofar as they increase the patent holder's reward, they encourage the patent system itself and therefore the incentive for R&D."[27]

Antitrust and IP are asymmetrical. Antitrust can forcibly separate someone's exclusive IP rights from his or her creative works, as Barnett shows occurred in the twentieth century. The scriptural creativity-ownership linkage does more to promote competition and improve consumer welfare than does an overzealous antitrust regime. The policies the Young Commission and the New Madison Approach advance fit more closely with biblical principles for human creativity and owning the fruits of one's labor than do policies empowering the government to deprive private property rights from those who invest themselves in inventive pursuits.

Second, Congress should **codify the 2019 joint policy statement on remedies available to standard-essential patents** subject to a RAND or FRAND licensing commitment.[28] This law should bind the FTC to this policy. It should vest in the DOJ Antitrust Division a veto over the FTC, mandating Justice to ensure the appropriate balance at the patent-antitrust nexus.

The 2019 SEP policy statement constitutes a facet of the New Madison Approach. Then-Assistant Attorney General Makan Delrahim explains SEP "owners cannot violate the antitrust laws by properly exercising the rights patents confer, such as seeking an injunction or refusing to license such a patent."[29] Affirming in statute the government must afford SEPs the full panoply of patent rights not only aligns with biblical creativity-ownership principles, it also promotes dynamic competition, which flows from innovation and improves consumer welfare.

Third, we must **repeal the Biden administration's antitrust initiatives to weaken patents**.[30] The 2021 executive order (E.O.) 14036, containing seventy-two wide-ranging orders and employing a "whole-of-government effort," seems more about scoring political points with particular special interests and advancing extreme ideological ends than promoting competition or rectifying objectively identified problems. The way the Biden FTC, PTO, NIST, and other agencies have carried them out bears witness to an ideology-over-substance intent. The agenda this E.O. laid out dramatically strengthens antitrust against IP (e.g., SEPs, Bayh-Dole, the Hatch-Waxman Act, and biotech patents). If this E.O.'s antitrust weaponization stands, patent rights will suffer for years as they did from the 1930s through the 1970s.

* * *

For more than thirty years, the United States has weakened its patent laws. The retreat characterizing American patenting 4.0 takes a toll on the United States and its people. It jeopardizes U.S. competitiveness, economic and technological benefits that inventors who fall victim to predatory patent infringers or government takings could have realized, and key elements of the U.S. patent system that formerly distinguished the American model.

Evidence shows secure, enforceable patents contribute to economic and technological gains. Stephen Haber has measured the connection between strong patent rights and economic growth. Haber reports, "There is nothing ambiguous about the resulting pattern: there are no wealthy countries with weak patent rights, and there are no poor countries with strong patent rights."[31] The only path available for the United States (or any other nation) to maximize its human innovative potential and thus its economic and technological progress is to restore the features that characterized American patenting 3.0, our golden age of patenting and invention. As C. S. Lewis suggests, we should backtrack to where we took a destructive path and get on the path leading to human flourishing through creativity and ownership.[32]

China is playing to win the technological leadership race.[33] Yet American attention is diverted by predatory infringers, including Big Tech behemoths, that game the U.S. patent system they have weakened. We must wake up and focus on recovering the creativity-ownership patent model, if the United States is to stand a chance in the global competitiveness contest.

Moving the needle demands we reconnect the creative act of invention with ownership of one's inventive works. To do so, our country must make significant reforms in patent and antitrust policy. Restoring access to permanent injunctions against patent infringers is top priority. Closely following are recovering broad patent subject-matter eligibility; reestablishing the private property status of patents and inventions; dismantling PTAB or imposing strict fairness, due process, and just compensation requirements on PTAB; and uprooting price-based Bayh-Dole march-in efforts. Adoption of other reforms, such as ending automatic eighteen-month publication of pending patent applications, restoring the grace period and first-to-invent, and repealing the prior-user loophole, would complement the big-ticket reforms and further strengthen the connection of creativity and ownership. Writing these patent policies into law would go a long way toward bringing about American patenting 5.0, a new golden age in which human flourishing thrives.

Patent reform is necessary, but that alone is unlikely to be sufficient. Policymakers must reset antitrust law so that government and courts give innovation's contribution of dynamic competition, creation of new markets, etc. their due. The law should require competition agencies to defer to the essential element of patents, the right to exclude. The controlling rule should operate by Jorde and Teece's observation "It is dynamic competition propelled by the introduction of new products and new processes that really counts [in enhancing economic welfare]."[34] New products and processes depend on patent exclusivity.

Codifying the New Madison Approach and the 2019 DOJ-PTO-NIST SEP policy statement, along with uprooting the Biden administration's aggressive antitrust policies that encumber IP rights, would recalibrate antitrust and promote invention through robust, secure patent rights that produce dynamic competition. These changes should prevent a return of a strong-antitrust, weak-IP regime like the one that arose in the twentieth century. That model distorted America's R&D funding and patenting and erected huge barriers against small entities with disruptive technologies. Otherwise, nimble startups with patented inventions might have displaced entrenched firms with the newcomers' superior technologies.

Getting back on the right road would help revive America's virtuous circle, an updated American patenting 3.0's innovation ecosystem. It made U.S. invention and commercialization the globe's undisputed innovation leader, coupled with remedies to contemporary challenges. Acting on Haber's conclusion that strong patents and strong economies go together (as do the opposites), this evidence-based policymaking—adopting the recommended patent and antitrust policy reforms discussed above—would launch American patenting 5.0. Our national innovation ecosystem would once again embody biblical creativity-ownership principles.

We would reinstate a regime fostering human flourishing because of its Common Grace foundation and humankind's divine image-bearing that includes God's communicable attributes, creativity and ownership.

Chapter 18

Conclusion: Reinventing by a Biblical Return

*"God says, in effect, 'Learn of the truth
that I have made in the external world.'"*
— Francis Schaeffer[1]

*"Well done, good and faithful servant. You have been faithful over a little;
I will set you over much. Enter into the joy of your master."*
— Matthew 25:21

Creativity and ownership, two foundational precepts, are attributes of the God of the universe. As we have seen, the opening words of Holy Scripture in Genesis underscore God's creativity: "In the beginning, God created the heavens and the earth." And Psalm 24:1 states God owns his creation: "The earth is the Lord's and the fullness thereof, the world and those who dwell therein."

The Creator God works and takes pleasure in his labor and in his creations. The Lord owns what he makes, his creation claimed as his private property.

In human beings, whom God created in his own image—distinct from all other creatures—he has endowed these same qualities, creativity and ownership. The Creation Mandate is humankind's divine calling to the dignity of work.

Distinguishing his and our creativity means our Heavenly Father can create from nothing, while we can only make something new from extant materials. Think of the abundance in the Garden of Eden. The work God called Adam and Eve to do there—to work and keep it—wasn't burdensome. In the state of innocence, their work and creative efforts were a source of unadulterated human flourishing. Adam owned what he made (temporally; everything ultimately belongs to the Creator). Adam quite literally was free to enjoy the fruits of his labor; he consumed the food his work produced.

After their fall from grace, Adam and Eve's work and creative endeavor became more taxing. Nevertheless, even after the Fall, God provides the means to survive and even to thrive. For example, God told the Israelites just before they entered the Promised Land it was a land of plenty with such raw materials as stones of iron and copper deposits (Deut. 8:9).

Invention and discovery are manifestations of our God-given creative quality. As we have seen, exercising our creative muscles involves effort, investing and expending our energy and resources to arrive at a viable solution. Inventing something that works and is commercially viable takes time, effort, and resources. Creative work isn't necessarily merely utilitarian. Whether creating useful art or fine art, works may have utility, beauty, or both.

Every human being has talents and abilities. And we each are responsible for our use of these gifts. Os Guinness says, "Our gifts are ultimately God's, and we are only 'stewards'—responsible for the prudent management of property that is not [ultimately] our own."[2] This encompasses responsibility for our creative efforts.

We can gain fulfillment beyond practical benefit or remuneration from creative endeavors. We're made to be creative. Being creative is good for us. Recall that human flourishing derives from using the skills and talents God has given us, doing what God has gifted us to do, thereby being what we're created to be.

Patents and intellectual property are means of securing inherent rights of ownership in a fallen world. We have a right to own the newly created property we each create, as we steward on the earth. As beings made in our Creator's image, we inherently possess this characteristic of property rights, just as God does. The Ten Commandments show this. Because property rights are inherent in people, including of our discoveries and creations, they have been illuminated by John Locke, secured in the U.S. Constitution by James Madison, and applied by inventors.

• • •

The combination of these two divine attributes, creativity and ownership, is an aspect of Common Grace. God provides for his creation through the creative endeavors of human beings. Leonardo da Vinci, Thomas Edison, the Wright Brothers, Jonas Salk, Hedy Lamarr, Jack Kilby and Robert Noyce, Samuel Morse, Jennifer Doudna and Emmanuelle Charpentier, George Washington Carver, Elon Musk, and others, some of them Christians and others not, each has a role in the Lord's providence, advancing the progress of science and the arts. Our patent system has been key in this aspect of Common Grace that benefits people around the world, blessing both present and future generations.

Thanks be to God for this marvelous model!

Common Grace ennobles the meaning and value of human creative endeavors. Our bearing God's image gives our creativity worth. The author and Christian Madeleine L'Engle observes how "In the beginning, God created the heavens and the earth" contains "a truth that cuts across barriers of time and space." She notes that "almost all of the best children's books [do too]," and "even the most straightforward tales say far more than they seem to mean on the surface."[3] This part of human intention and its likeness to our Creator's multilayered meaning and purpose reflect our imago Dei nature.

A Legacy from Our Creating and Owning

People tend to want to leave their mark on the world. As we get further into the years of our lives, people wonder if their life will have made a difference. It's natural to want to leave a legacy. Inventors and creators are no different in this. Some people leave financial assets to an alma mater or a cultural, scientific, or charitable institution. Others leave artistic, technological, literary, or other possessions to human or institutional beneficiaries.

Humans are finite. Our lives are finite. Our Maker "remembers that we are dust." Psalm 103 continues: "As for man, his days are like grass; he flourishes like a flower of the field; for the wind passes over it, and it is gone, and its place knows it no more" (vv. 15–16). Every human being's life on earth comes to an end. All the stuff we accumulate over our lifetimes we cannot take with us into eternity. Only our souls live on until joined with our resurrected bodies.

For creators and inventors, their creative and inventive works may be preserved. Those creations or a newer, improved iteration of a work's essence or its essential function may remain in use into the future. Our smartphones today fulfill a range of functions, one of them being audio communications. An Apple iPhone or a Samsung Galaxy, through its telephone feature, delivers the same function as Alexander Graham Bell's telephone or the rotary phone of the twentieth century: a voice conversation between two people. And text messaging owes much to the telegraph. SMS is a new and improved way of sending and receiving instantaneous communications by electronic means—the essential function of the telegraph.

In this manner, the creative works of long-passed inventors, musicians, artists, architects, and other creators continue to benefit people into the future. Such creative successes may help keep the memory of an inventor or creator in the public's mind.

● ● ●

The fleeting existence of a human being, likened to grass or a flower that the wind blows and the sun scorches to its demise, is remembered in the future only by some legacy. Underlying such remembrance is achievement by God-given creative ability coupled with property rights in the creative output of that individual. Creating previously nonexistent property and having the economic freedom to own that property are necessary for success for the individual, benefits for society, and the prospects of leaving a legacy.

Our discussion above regarding humankind's Fall and the Creation Mandate provides context for consideration of obtaining a legacy of creative works and the fruits thereof. First, science and the useful arts provide the tools for achieving progress on earth in repairing the consequences of the Fall. Francis Bacon, the sixteenth- and seventeenth-century English scientist, writes: "For man, by the fall, lost at once his state of innocence, and his empire over creation, both of which can be partially recovered even in this life, the first by religion and faith, the second by the arts and sciences."[4] Nancy Pearcey remarks, "By *arts* Bacon meant the technical arts, and his point was that the scientific study of nature, applied through technology, could be used to reverse the effects of the fall."[5] That is, the tools and talents of inventors and scientists can serve the purpose of advancing the human condition and caring for creation.

Bacon seeks to persuade humanity to strive for the highest type of ambition. He discusses the "vulgar and degenerate" ambition to accrue power and control for oneself over his own country and people. Second, he addresses the more dignified but still covetous ambition to expand power and dominion over other lands.

Finally, Bacon lauds the betterment of humanity. "If a man endeavor to establish and extend the power and dominion of the human race itself over the universe, his ambition (if ambition it can be called) is without doubt both a more wholesome thing and a more noble than the other two. Now the empire of man over things depends wholly on the arts and the sciences, for nature is only to be commanded by obeying her."[6]

While the first two ambitions remain at large, Bacon might take comfort in the extraordinary progress from the sciences and the useful arts mankind has achieved. He would surely recognize the source of, and attribute these fruits of human labor to, the biblical combination of creativity and ownership. The creature made in God's image is working out the Creation Mandate under God's Common Grace, which blesses humanity.

* * *

Kentucky Congressman Thomas Massie is an inventor, patent owner, and entrepreneur whose fascination with arms and hands set him on a course to push the technological envelope with haptic computerization and artificial intelligence connected to robotic hands. He echoes Bacon's third, noblest ambition: "extend[ing] the power and dominion of the human race itself over the universe."

Rep. Massie says it's "overstate[d,] the degree to which politicians are going to be responsible for whether society advances or not. I think most of the improvements in our lives have come from entrepreneurs and inventors with the long-term improvements."[7]

Massie speaks as one who has lived a life lesson. He knows the power of the arts and sciences, exercised by millions and millions of individuals applying them to real-world challenges in multiple scientific and technological fields, will accomplish far more good for humankind than will political bodies. For statesmen, the coins of the realm are speech and debate, coalitions and group decisions.

"The human condition over time is improved by adventurous entrepreneurs who invest in new ideas, and I'm living vicariously through those people," Massie, the MIT-trained computer engineer, says.

> And when people ask me, will the next generation be worse off or better off than the generation we're in right now, I say it's almost certainly going to be better off. But not because of politicians. It's going to be because of the inventors and entrepreneurs and disruptors.
>
> And our obligation as politicians is not to screw up that environment, not to make it so hard that they can't improve our lives, when politicians are not the ones who are going to improve our lives. It's going to be inventors.[8]

Bacon might smile at this statement.

* * *

Another facet of Bacon's third ambition relates to God's model for democratized invention and innovation. As we know, creativity and ownership are common attributes in all people. It's usually people driven by the two baser ambitions who erect artificial barriers to keep others from engaging—or at least to benefit from—these attributes in the innovation prospect.

Nobel economist Edmund Phelps's modern economic model, based on the fact "capitalism creates innovation," supports Haber's findings about the correlation of strong patents and strong economies. It also empirically validates the Founders' democratized patenting model, rewarding merit over cronies.

Phelps finds, "Countries with more inclusive institutions, irrespective of any other advantages or handicaps they might have, consistently have better records of long-term economic growth."[9] That is, democratized, merit-based patent and IP rights grow an economic pie faster because everybody has a fair shot at being rewarded for their inventive creativity and having secure title to the fruits of their labors, if they have the grit, ability, and dedication to solve the problem at hand. Such democratized participation and merit-based reward "enlist a larger share of the national IQ, so to speak, into the scramble for innovation."[10]

The Cade Museum's "Inventivity" initiative aligns with Phelps's and Haber's scholarship and God's creativity-ownership model under Common Grace. "Gatorade was saved by a housewife, who didn't even think of herself as an inventor," Phoebe Miles, daughter of Gatorade's inventor, says. "My mom didn't think of herself that way. Which proves . . . you don't have to be an inventor to have an inventive mindset."[11]

Miles and Inventivity hold to the Common Grace solution to individual, national, and humanity's improvement through human creative effort by way of the private property right of exclusive ownership to one's inventive solution. "We think it's essential to our nation," Miles says. "Why? Because cultivating an inventive mindset develops a culture of innovation. And a culture of innovation is the difference between thriving and surviving."[12] This is Phelps, phrased differently.

* * *

Whether the way of discussing the creativity-ownership model resonates most as phrased by Bacon, Massie, Phelps, or Miles, they all are talking about the same thing. God's Common-Grace means of humans applying their own sets of talents in creative or inventive endeavor is the solution to humanity's problems. Creativity and ownership must be equal partners in this model for it to work. This places us at forks in the road.

A Time for Choosing

Postmodern humankind holds to an untenable dichotomy of extraordinary technological advancement on one side and rejection of the objective, Judeo-Christian worldview, which long informed the West's understanding of the true, the good, and the beautiful, on the other side. The centrifugal effect of this dichotomy is "reality" becomes surreal and civil society disintegrates. The Western, particularly American, rush toward hyper-individualism occurring simultaneously with the quickening pace of technological progress exemplifies this dichotomy.

The dichotomy comes down to a contest between worldviews. Several choices confront us, though there are essentially two dueling worldviews. One holds that God is at the center, the King of kings, the Creator and Owner of everything, and thus his image-bearers, we humans, owe him our all. The other worldview, though it comes in many variations, puts human beings at the center of the universe; it presumes to displace God from his throne.

Whether one consciously engages in the philosophical or spiritual debate over or the battle of worldviews, one feature of the contest involves a choice of intellectual property (and property rights generally) edifice. Everyone has a dog in this fight because everyone benefits or loses from the choice made, individually and as a body politic.

• • •

First, as a nation we must choose a course of public policy. Our options at this point are clear: We recover a regime of private property rights that secures property rights in one's inventive and creative works, paired with antitrust laws whose reach is constrained where the exercise of intellectual property is concerned; or we continue down the weak-property-rights, aggressive-antitrust course that undermines the creativity-ownership framework.

Getting us back on the right road of patent policy would help revive America's virtuous circle. American patenting 3.0's innovation ecosystem made U.S. invention and commercialization the world's undisputed innovation leader. We should return to this model with updated remedies that address contemporary challenges. Once we adopt this option, our officials must vigorously promote the robust-patents, modest-antitrust model to foreign commercial-partner nations on behalf of U.S. innovators active in their markets and fellow member countries of international bodies such as the World Intellectual Property Organization and the World Trade Organization.

Acting on Haber's conclusion that strong patents and strong economies go together (as do their opposites), evidence-based policymaking—

adopting the recommended patent and antitrust policy reforms discussed previously—would launch American patenting 5.0. Our national innovation ecosystem would once again employ biblical creativity-ownership principles. Reinstating this regime will foster human flourishing because of its Common Grace foundation and humankind's divine image-bearing with God's communicable attributes, including creativity and ownership in combination.

On the other hand, if the United States continues down the road of weakening IP rights and strengthening antitrust (or other regulatory means of "taking" private property), the consequences are dire. In a fallen world in which work is harder and sinful human beings lie, cheat, and steal from others, human flourishing withers and the default setting of poverty takes root and blossoms (as it followed Adam's fall from grace in the Garden or is found in any socialistic or communistic regime).

That is, leaving private property rights unprotected disincentivizes creativity, work, and innovation. The default is poverty; prosperity is not the norm. Wealth is created only when property rights are secure. A weak-IP regime turns private property into the tragedy of the commons. What inventor, creator, or IP owner will develop or improve creative works if it's turned into common property, placed in the public domain by weak property rights?

This should be an easy choice. As the American invention history demonstrates, aligning our IP framework with biblical creativity-ownership principles is like cutting wood with the grain; it's smooth. What our IP-antitrust regime has slipped into cuts against the grain of the Creator's sovereign design for our well-being.

* * *

Second, inventors, creators, scientists, and engineers face other choices. These relate to the legacy question discussed above: Will their lives make a difference in this world? How will they be remembered? This is ultimately a moral choice or set of choices.

Creative and inventive individuals can discover, invent, create, and commercialize for good or for evil. They may not think of their decisions made along the creative path in this manner because, generally speaking, technology is neutral. Often, a creative work's or an invention's use dictates its moral or immoral nature. For instance, the World Wide Web disseminates the Gospel of Jesus Christ, the scientific laws of thermodynamics, instructions for fixing a leaky air conditioner pipe, copyright-infringing movies and music, vile prurient contents, lies and propaganda, hate-filled diatribes, and innumerable cat videos.

Likewise, CRISPR gene-editing technology has been used both morally and immorally.[13] Artificial intelligence, biotechnology, and quantum computing could be employed for both good and evil ends, as technological advancements through the ages have been.

Human agency determines an invention's or creation's purpose and usage—not necessarily the maker. Immoral use may be the inventor's intent or the choice of a criminal, a tyrant, an ambition-driven climber, or a sorely misguided zealot. Gifted German doctors and scientists who devoted their talents to aiding and abetting the Nazi regime; skilled Chinese engineers and inventors who labored on behalf of Mao's or serve Xi's Communist, totalitarian state; scientists and technologists who put various neutral inventions to serving the purposes of the Soviet Union: These and their ilk are remembered in infamy. In contrast, Edison, the Wright Brothers, Alexander Graham Bell, and other inventors and creators are remembered favorably for the beneficial works they conceived and brought to fruition.

Questions with moral implications will arise in a wide range of contexts and stages of one's creativity and ownership. Better to anticipate them and consider them carefully along the way than to be surprised long afterward, when the consequences have mounted dramatically.

● ● ●

Third, another set of choices relates to the second set just considered, only these have both deeper, more personal and broader, more philosophical implications. These choices concern one's worldview, the place technological innovation holds in one's heart and mind, and whether one's fate and legacy depend upon creativity and ownership in themselves or something else.

In 1993, Christian intellectual Os Guinness wrote, "Reliance on technology has become almost an article of faith in America."[14] Guinness acknowledges "Yankee ingenuity and know-how have already shown themselves inventive and flexible enough to steer America through many crises."[15] However, he says America has lost her moral compass. Our nation has replaced the Judeo-Christian moral structure with a creative-inventive idol, in which skilled humans play an outsized role. Guinness notes, "American faith in technique is Protestantism shorn of all transcendence whatsoever."[16]

America's (and other techno-fixated nations') "counting on technology to compensate for the loss of morality" is misplaced faith. Technological solutions sometimes fail or "[create] a host of new dilemmas whose solution requires morality even more urgently." Moreover, "technology cannot take into account the essential irrationality of human evil and therefore can never substitute adequately for traditional morality."[17]

Guinness is saying the United States even thirty-plus years ago made technology our society's collective idol. "In technology we trust," he quips. In displacing the Creator God with devices of our own making, we, like the builders of the Tower of Babel, shift from applying the divine attribute of creativity in the work to which he has called us for obtaining rightly motivated goals to using these abilities to glorify ourselves. We redirect God-endowed ownership to selfish and avaricious ends. Or, in the other extreme, we hollow out ownership, transforming it into "public property." That results in the tragedy of the commons, where the state—ostensibly everyone—"owns" property, which largely goes unimproved, unused or ill-used.

The picture Guinness paints communicates the same thing as Psalm 115: "Their idols are silver and gold, the work of human hands. They have mouths, but do not speak; eyes, but do not see. They have ears, but do not hear; noses, but do not smell. They have hands, but do not feel; feet, but do not walk; and they do not make a sound in their throat. *Those who make them become like them; so do all who trust in them*" (vv. 4–8, emphasis added).

Self-centeredness in and presumption regarding our own abilities and fruits make us our own idols. God warned the Israelites against this sin before they entered the Promised Land, saying, "Take care lest you forget the LORD your God . . . who brought you out of the land of Egypt, out of the house of slavery. . . . Beware lest you say in your heart, '*My power* and the might of *my hand* have gotten me this wealth'" (Deut. 8:11, 14, 17, emphasis added).

We must keep in mind the source of humanity's creativity and ownership characteristics. The infinite-personal God of the Bible, who created human beings in his image, is eternal. He is sovereign over all the universe and Heaven. He made creativity and ownership, and he graciously shares these attributes with his highest creature. Congressman Massie explains that human beings are "endowed by God with the ability to change our environment, not just to be participants or victims of our environment. We are able to change our environment and make tools that let us change our environment even more."[18]

As the only rational beings in all creation, humans have these and other inherent qualities, and thus have inherent dignity. We forget this, as well as the fact we too have an eternal existence. Schaeffer writes: "People today are trying to hang on to the dignity of man, but they do not know how to, because they have lost the truth that man is made in the image of God."[19]

By endowing communicable attributes such as rationality, creativity, and ownership in his image-bearing creatures, God gives humans the faculties, talents, and abilities, and the incentives to change our environment. That is, we possess what we need to carry out the work of the Creation Mandate. And those people who don't put their faith in Jesus Christ for eternal salvation, in God's sovereignty, may also participate in the work that brings individuals and the human race human flourishing in this world. All individuals may enjoy the blessings of God's Common Grace—a testament to human dignity because of imago Dei.

This discussion invites us to consider our worldview. Is it more or less Judeo-Christian, egocentric, or humanistic? This section also opens the door to consider whether creativity, ownership, technology, or the divine Maker of all this is our actual object of worship.

Thinking these things through may present new choices to make. This thought process and the choices that follow bear upon the question of our legacy, our life's mark. I pray it will be a fruitful exercise, renewing and encouraging in the process. This time for choosing may reset and reinvigorate our creative juices, our aspirations, our vision, and our priorities.

Conclusion: Reinventing by a Biblical Return | 265

• • •

This book aims to illuminate a number of concepts, terms, and their applications to real-world elements associated with human creativity and the ownership of creative output. These are vitally important to know, understand, and revive because for centuries they were well known throughout American society; they informed the American property rights and intellectual property systems, laws, policies, and practices from the Founding era into, and largely through, the twentieth century. The Judeo-Christian ethic characterizing our nation from colonial days through the mid-twentieth century effectively made these presuppositions of the foundational American patenting model.

Therefore, reintroducing these terms, concepts, and applications into the American conversation about creativity, ownership, IP, technology, and human flourishing will help us improve our IP system. We'll be equipped to consider these modern issues from a sound perspective that has long served our nation well.

These components of the biblical creativity-ownership framework are vitally important to human flourishing. Understanding the crucial link of ownership of one's creative output to human flourishing, for both individuals and societies, we can restore crucial elements of the Founders' system. Their culture's Reformation and Great Awakening heritage informed the vision of using private property rights as incentives to be inventive or creative, supercharging human progress and wealth creation.

Alas, secularization's rise along with widespread biblical illiteracy in the postmodern United States has brought cultural amnesia. This book is the wake-up call. And it delivers the information that could empower every American, from grade-school robotics contestants to engineers and scientists to researchers to commercialization specialists to policymakers to IP attorneys to venture capitalists to garage inventors to movie moguls to musicians.

Consider Francis Schaeffer's words about the untapped power we, as individuals and as a nation, ignore through our unfaithfulness. Thus, we miss out on many unrealized blessings:

> The Lord will not honor with power the way of unfaith in His children because it does not give Him the honor. He is left out. That is true in Christian activities, in missionary work, in evangelism, in anything you name. . . . Who can do more? We with our own energy and wisdom, or the God who created Heaven and earth and who can work in space-time history with a power which none of us has? God exists. And if we through faith stay in the Bible-believing chair moment by moment in practice, and do not move into the chair of unfaith, Christ will bring forth His fruit through us. The fruit will differ with each of us, but it will be His fruit.[20]

I hope this book has reminded your heart, soul, and mind that you, with your abilities, interests, creativity, and inherent ownership, are made in God's image, and these are his gifts to you so you may thrive in human flourishing.

Acknowledgments

This book owes its existence to many people whose help I deeply appreciate for their part in bringing it to fruition. First, Diane Truitt invited me to speak at her church on the topic of patents and invention from a biblical perspective. She, along with Gary Bennett and Jim Haines, arranged for my speaking at one of First Baptist Dallas's Discipleship University classes. I appreciate their getting this ball rolling. Thanks to Rebekah Gantner, Dave Murray, Gabe Neville, and Dick Patten for letting me practice my speech on them and for their feedback. Ed Martin gave me a speaking session at a Phyllis Schlafly Eagles conference, honing it further.

Thann Bennett explained the business side of writing and getting a book published and provided instructive feedback on my book proposal. Eden Gordon Hill filled me in on the promotion-publicity aspect of marketing a book. Myra Berry timely gave me several providentially selected sources for this book. Charlie Sauer shared a sample book proposal.

I'm grateful to five inventors/creators who allowed me to interview them: Susie Armstrong, a world-class technologist whose work in computer communications protocols at Xerox and mobile computing at Qualcomm have benefitted billions of people; Claire Kendall, a talented artist who displays technical excellence while communicating depth; U.S. Rep. Thomas Massie, a leader in the House Intellectual Property Subcommittee whose principled advocacy on patent policy is informed by multifaceted, real-world experience; John McCorkle, inventor, patent owner, and entrepreneur who is globally recognized as a leader and pioneer in ultrawide bandwidth engineering; and Cephus Simmons, a medical professional who saw a clinical need, conceived a solution, and invented it.

Thanks to Phoebe Miles, a daughter of Gatorade's lead inventor and a founder of the Cade Museum for Creativity and Invention, who has been supportive of this effort from the beginning. I appreciate her encouragement and collaboration. I'm grateful to Faith & Law for the opportunity to share new material, incorporated and expanded in this book, before an attentive, engaged audience.

The quality of this book is enhanced by the able assistance of librarians at the Library of Congress and by Paul Israel, director and general editor of the Thomas A. Edison Papers at Rutgers University. Moreover, I greatly appreciate the expert eyes of Judge Paul Michel (ret.), former chief judge of the Federal Circuit, on draft chapters. Likewise, American patent law and history expert Adam Mossoff reviewed several chapters. Any errors are my own.

I'm in debt to my friend Tim Goeglein for his encouragement, sharing his insights into book writing and publishing, and more. Gary Terashita, Miko Griffin, and Erin Ashley of Fidelis Publishing have been a pleasure to work with on this project.

Invaluable throughout this project has been my daughter, MacKenzie, who's performed a great deal of research, proofread, offered editorial suggestions, fact-checked, and more. I also thank the friends who kept me and my book project in their prayers. Special thanks to my wife, Linda, for putting up with my spending much time on this project. Above all, thanks to the Lord God, the King of Creators, who owns all he has made. His bringing this project to me has produced flourishing through these efforts.

List of Photographs and Images

Chapter 5:

Photograph by John T. Daniels, "Wright Brothers, First flight, 120 feet in 12 seconds, 10:35 a.m.; Kitty Hawk, North Carolina," December 17, 1903, reproduction no. LC-DIG-ppprs-00626, https://www.loc.gov/rr/print/res/269_wri.html, Library of Congress Prints and Photographs Division, Washington, DC. No known restrictions on publication.

Chapter 14:

Anon., *United States Patent Office, Washington, D. C.*, Washington News Co. lithograph, ca. 1915–1930, https://www.si.edu/object/united-states-patent-office-washington-d-c:npg_NPG.POB144 National Portrait Gallery, Smithsonian Institution, Washington, DC. Public domain, Creative Commons.

Christian Schussele, *Men of Progress*, 1862, https://www.si.edu/object/men-progress:npg_NPG.65.60, National Portrait Gallery, Smithsonian Institution, Washington, DC. Public domain, Creative Commons.

Chapter 15:

Anon., *Model Hall of the United States Patent Office, Washington, D. C.*, lithograph, n.d., https://www.si.edu/object/model-hall-united-states-patent-office-washington-d-c:saam_1967.108, Smithsonian American Art Museum, Washington, DC. Public domain, Creative Commons.

Anon., "Telegraph Receiver" invented by Samuel F. B. Morse, patent no. 6420, n.d., https://www.si.edu/object/telegraph-register:nmah_706500, Smithsonian National Museum of American History, Washington, DC. Public domain, Creative Commons.

Chapter 16:

Photograph by MacKenzie R. Edwards, "Current U.S. Patent & Trademark Office, 2003–present, Alexandria, VA," 2023, courtesy of MacKenzie R. Edwards.

Endnotes

Chapter 1: A Peek into Human Creativity and Its Divine Source

1. Phoebe Miles, "To Invent Is Divine: How Strong Patent Rights Help Us to Reflect God's Innovative Nature," with James Edwards, at Faith & Law, *Friday Forum*, July 8, 2022, 2172 Rayburn House Office Building, Washington, DC, video, 01:10, https://faithandlaw.org/resources/2022/05/to-invent-is-divine/.

2. T. R. Reid, *The Chip: How Two Americans Invented the Microchip and Launched a Revolution*, rev. ed. (Random House, 2001), 72–77.

3. An electric circuit is the unbroken route that electricity follows. Circuits have four essential components: A *resistor* regulates the electrical flow; a *capacitor* collects electrical energy and releases it in the volume needed; a *diode* is like a door that is open or closed, either allowing electricity to flow or blocking electrical current; a *transistor* uses semiconductor material to control electricity's flow. Reid likens these components to a nozzle, a sponge, a dam, and a faucet, respectively. See Reid, *The Chip*, 13–14.

4. Reid, *The Chip*, 87–95.

5. "Integrated Circuits: 1958: Invention of the Integrated Circuit," PBS, https://www.pbs.org/transistor/background1/events/icinv.html (accessed July 15, 2024).

6. Phill Parker, "Apollo and the Integrated Circuit" (December 2003), in *Apollo Lunar Surface Journal*, Ken Glover, ed., and Eric M. Jones, founder and ed. emeritus, https://www.hq.nasa.gov/alsj/apollo-ic.html.

7. Jack Kilby, "The Integrated Circuit's Early History," *Proceedings of the IEEE* 88, no. 1 (January 2000): 111.

8. "Revolutionizing the Mobile Internet," *Qronicles: A Compilation of Qualcomm Inventions*, Qualcomm, 2016, 12–13; *Connectivity Changes Everything*, 2018–2020 Qualcomm Exhibit, National Inventors Hall of Fame, U.S. Patent and Trademark Office, 4–8.

9. See Cornelius Plantinga, Jr., *Not the Way It's Supposed to Be: A Breviary of Sin* (Wm. B. Eerdmans, 1995).

10. Senate, report no. 1979, 82nd Cong., 2nd sess., 1952, https://www.ipmall.law.unh.edu/sites/default/files/hosted_resources/lipa/patents/Senate_Report_No_1979.pdf (accessed July 15, 2024): 4. The clause is cited in *Diamond v. Chakrabarty*, 447 U.S. (1980): 303, 308.

Chapter 2: Making Things and the Fruits of Your Labor

1. "Irving Berlin Biography," Biography.com Editors, A&E, https://www.biography.com/musicians/irving-berlin (accessed July 13, 2024).
2. "Michelangelo," Encyclopaedia Britannica, https://www.britannica.com/biography/Michelangelo (accessed July 14, 2024).
3. "Leonardo da Vinci Biography," Biography.com Editors, A&E, https://www.biography.com/artist/leonardo-da-vinci (accessed July 14, 2024).
4. George Pratt, "Hallelujah! The Story of Handel's Messiah," *BBC Music Magazine*, December 8, 2018, http://www.classical-music.com/article/hallelujah-story-handel-s-messiah.
5. "William Shakespeare Biography," Biography.com Editors, A&E, https://www.biography.com/authors-writers/william-shakespeare (accessed July 13, 2024).
6. Orville Wright, "How We Made the First Flight," *Flying*, 1913, repr. 2003, MN Dept. of Transportation, 14.
7. Article I, Section 8, Clause 8.
8. W. Bernard Carlson, "Documenting Invention: Developing a Flow Model of Invention," Lemelson Center for the Study of Invention and Innovation, Smithsonian Institution, September 1, 2007, https://invention.si.edu/documenting-invention-developing-flow-model-invention.
9. Carlson, "Documenting Invention."
10. Phoebe Miles, personal communications with author, September 4, 2020, and August 23, 2024.
11. "Our Heritage: Changing the Game Since 1965," Gatorade, https://www.gatorade.com.au/our-heritage (accessed July 15, 2024).
12. Carlson, "Documenting Invention."
13. Fred Howard, *Wilbur and Orville: A Biography of the Wright Brothers* (Dover, 1998), 15–16.
14. Robert Yonover and Ellie Crowe, *Hardcore Inventing: Invent, Protect, Promote, and Profit from Your Inventions* (Skyhorse, 2009), 13–14.
15. Howard, *Wilbur and Orville*, 11.
16. Justin Taylor, "An Interview with Andy Crouch about the Idol and Gift of Power," Gospel Coalition, September 30, 2013, https://www.thegospelcoalition.org/blogs/justin-taylor/an-interview-with-andy-crouch-about-the-idol-and-gift-of-power/.
17. John Dickinson, *Letters from a Farmer in Pennsylvania to the Inhabitants of the British Colonies*, letter 5, 46n (n.p., 1774), https://history.delaware.gov/john-dickinson-plantation/dickinsonletters/pennsylvania-farmer-letters/ (accessed July 13, 2024), 48–49.
18. Harold Lindsell, *Free Enterprise: A Judeo-Christian Defense* (Tyndale House, 1982), 14.
19. Thayer Watkins, "The Economic Collapse of the Soviet Union," San Jose State Univ., https://www.sjsu.edu/faculty/watkins/sovietcollapse.htm (accessed July 12, 2024).
20. Wendy H. Schacht, "The Bayh-Dole Act: Issues in Patent Policy and the Commercialization of Technology," *Report for Congress* RL32076 (Washington: Congressional Research Service, December 3, 2012), https://sgp.fas.org/crs/misc/RL32076

.pdf. Also see Lori Pressman, Mark Planting, Jennifer Bond, Robert Yuskavage, and Carol Moylan, "The Economic Contribution of University/Nonprofit Inventions in the United States: 1996–2017," Biotechnology Innovation Organization (BIO) and AUTM, June 5, 2019, https://autm.net/AUTM/media/About-Tech-Transfer/Documents/Economic_Contribution_Report_BIO_AUTM_JUN2019_web.pdf).

21. U.S. Constitution, Article I, Section 8, Clause 8.

22. "Second Lecture on Discoveries and Inventions," February 11, 1859, *Collected Works of Abraham Lincoln*, vol. 3 (August 21, 1858–March 4, 1860), 363. Univ. of Michigan Library Digital Collections, https://name.umdl.umich.edu/lincoln3.

Chapter 3: God the Creator

1. Matthew Henry, *A Commentary on the Holy Bible*, illustrated ed., vol. 1 (W. P. Blessing, n.d.), 2.

2. The creeds, confessions, and catechisms cited are from appendices of R. C. Sproul, *Reformation Study Bible*.

Chapter 4: Made in His Image: Human Creativity

1. Vernon Blackburn, "Wolfgang Amadeus Mozart," in *Music Lovers' Encyclopedia*, Rupert Hughes, comp., Deems Taylor and Russell Kerr, eds. (Garden City, 1947), 508.

2. *Intellect* refers to knowledge and wisdom. The human soul is comprised of intellect, or rationality, and will. These faculties each reflect qualities of God and are aspects of humanity's bearing of the divine image.

3. Matthew Henry, *A Commentary on the Holy Bible*, illustrated ed., vol. 1 (W. P. Blessing, n. d.), 7.

4. Francis A. Schaeffer, *Genesis in Space and Time*, in *The Complete Works of Francis A. Schaeffer: A Christian Worldview*, vol. 2 (Crossway, 1982), 33.

5. Henry, *Commentary*, vol. 1, 6.

6. For a fuller discussion of human beings' bearing God's image, see Louis Berkhof, "Man as the Image of God," in *Systematic Theology*, expanded ed. (Banner of Truth, 2021), 196–205.

7. Berkhof, *Systematic Theology*, 199–200.

8. Schaeffer, *Genesis in Space and Time*, 32.

9. Schaeffer, *Genesis in Space and Time*, 32. Also, see Richard D. Phillips, *The Masculine Mandate: God's Calling to Men* (Ligonier, 2010), and Andy Crouch, *Culture Making: Recovering Our Creative Calling* (InterVarsity, 2008).

10. Schaeffer, *Genesis in Space and Time*, 41.

11. Jonathan Master, *Growing in Grace: Becoming More Like Jesus* (Banner of Truth, 2020), 17–19.

12. Phillips, *Masculine Mandate*, 18, 49.

13. Phillips, *Masculine Mandate*, 19–21.

14. Crouch, *Culture Making*, 108.

15. Alison Gopnik, "What AI Still Doesn't Know How to Do," *Wall Street Journal*, July 16, 2022. See Andy Kessler, "No, AI Machines Can't Think," Inside View, *Wall Street Journal*, January 8, 2024.

16. Crouch, *Culture Making*, 104.
17. Crouch, *Culture Making*, 104.
18. Betty Alexandra Toole, *Ada, the Enchantress of Numbers: A Selection from the Letters of Lord Byron's Daughter and Her Description of the First Computer* (Strawberry, 1992), 136.
19. Toole, *Ada*, 137.
20. Letter from Alexander Graham Bell to Alice Jennings, February 12, 1872, Alexander Graham Bell Family Papers, Library of Congress, https://www.loc.gov/collections/alexander-graham-bell-papers/?q="god+has+strewn+our+paths+with+wonders" (accessed July 22, 2024).
21. Yogi Berra, *The Yogi Book* (Workman, 1998), 123.
22. Crouch, *Culture Making*, 106.
23. Crouch, *Culture Making*, 106.
24. See Schaeffer, *Genesis in Space and Time*, 22–26.
25. Cornelius Plantinga, Jr., *Not the Way It's Supposed to Be: A Breviary of Sin* (Wm. B. Eerdmans, 1995), 29.
26. Crouch, *Culture Making*, 107.
27. Crouch, *Culture Making*, 105.
28. Crouch, *Culture Making*, 105.
29. Crouch, *Culture Making*, 105.
30. Allysia Finley, "Electricity Is the New Medical Miracle," Life Science, *Wall Street Journal*, July 22, 2022.
31. Walter Isaacson, *The Innovators: How a Group of Hackers, Geniuses, and Geeks Created the Digital Revolution* (Simon & Schuster, 2014), 20.
32. Fred Howard, *Wilbur and Orville: A Biography of the Wright Brothers* (Dover, 1998), 11.
33. See Howard, *Wilbur and Orville* and David McCullough, *The Wright Brothers* (Simon & Schuster, 2015).

Chapter 5: Invention at Work

1. William J. Federer, *George Washington Carver: His Life and Faith in His Own Words* (Amerisearch, 2003), 54.
2. J. Presper Eckert, oral history interview with Nancy B. Stern, Charles Babbage Institute, Univ. of Minnesota, Blue Bell, PA, October 28, 1977, https://hdl.handle.net/11299/107275 (accessed July 12, 2024).
3. The origin of this saying comes loosely from Plato's *Republic*: "Your genius will not be allotted to you, but you will choose your genius . . ." Plato, *Republic*, Benjamin Jowett, trans. (Barnes & Noble, 2004), 347. See "Necessity is the mother of invention," Phrases.org.uk, Gary Martin, https://www.phrases.org.uk/meanings/necessity-is-the-mother-of-invention.html (accessed July 13, 2024).
4. "Dean Kamen," *Encyclopaedia Britannica*, https://www.britannica.com/biography/Dean-Kamen (accessed July 6, 2024).
5. "The Man Behind a Billion Connections," Qualcomm, https://www.qualcomm.com/research/stories/man-behind-billion-connections (accessed July 6, 2024).
6. Fred Howard, *Wilbur and Orville: A Biography of the Wright Brothers* (Dover, 1998), 64.

7. Howard, *Wilbur and Orville*, 64.
8. Howard, *Wilbur and Orville*, 67.
9. Howard, *Wilbur and Orville*, 73.
10. Howard, *Wilbur and Orville*, 33.
11. Howard, *Wilbur and Orville*, 34.
12. Howard, *Wilbur and Orville*, 137.
13. For more information about the process of invention, see Robert Yonover and Ellie Crowe, *Hardcore Inventing: Invent, Protect, Promote, and Profit from Your Inventions* (Skyhorse, 2009), chaps. 1–8; and Charles B. McGough, *Great Invention! Now What?: Evaluate, Patent, Trademark, and License Your New Invention* (Self-Counsel, 2014), chaps. 1-6. More examples of invention may be found throughout Fred Howard, *Wilbur and Orville*; Jill Jonnes, *Empires of Light: Edison, Tesla, Westinghouse, and the Race to Electrify the World* (Random House, 2004); Michelle Malkin, *Who Built That: Awe-Inspiring Stories of American Tinkerpreneurs* (New York: Threshold Editions/Mercury Ink, 2015); and T. R. Reid, *The Chip: How Two Americans Invented the Microchip and Launched a Revolution*, rev. ed. (Random House, 2001).
14. Frank Dyer and T. C. Martin, *Edison: His Life and Inventions*, https://babel.hathitrust.org/cgi/pt?id=mdp.39015014861382&view=1up&seq=167, 607. Paul Israel, director and general editor at Thomas A. Edison Papers, Rutgers Univ., points out this famous quote in the 1910 official biography actually blends elements from other reported Edison statements, such as a 98 percent-2 percent version in "The Anecdotal Side of Edison," *Ladies Home Journal* (April 1898): 7–8, https://edisondigital.rutgers.edu/document/SC98015A; Francis Arthur Jones, *Thomas Alva Edison: Sixty Years of an Inventor's Life* (1907), https://babel.hathitrust.org/cgi/pt?id=mdp.39015004535103&seq=421; and a most interesting marginal note by Edison on a letter from Mark H. C. Spiers, February 1, 1915, that "they attribute this saying to me but I did not [cross out] I cannot remember that I ever said it," https://edisondigital.rutgers.edu/document/E1525AC.
15. Susan M. Armstrong, interview by author, audio recording via Skype, October 23, 2023.
16. Armstrong, personal interview.
17. U.S. Patent and Trademark Office, *Hearings on the Study of Underrepresented Classes Chasing Engineering and Science Success (SUCCESS) Act*, June 30, 2019, statement of Susie M. Armstrong, 3, https://www.uspto.gov/sites/default/files/documents/SUCCESSAct-Armstrong.pdf.
18. James R. Hagerty, "Inventor Dreamed Up Better Way to Print," *Wall Street Journal*, January 18, 2020.
19. James R. Hagerty, "Chemical Engineer Invented Gore-Tex," *Wall Street Journal*, September 19, 2020.
20. *Paul McCartney: A Life in Lyrics* podcast, season 2, episode 4, 15:34, "Yesterday."
21. A cam is an irregularly shaped item that, when attached to a part of a device and set in motion, causes an intended, noncircular pattern, such as an ellipse.
22. Charles R. Morris, *The Tycoons: How Andrew Carnegie, John D. Rockefeller, Jay Gould, and J. P. Morgan Invented the American Supereconomy* (Henry Holt, 2005), 34, 35.

23. Eckert, oral history interview, Charles Babbage Institute.
24. Jonnes, *Empires of Light*, 39.
25. Jonnes, *Empires of Light*, 40.
26. "James P. Allison, Biographical," Nobelprize.org, https://www.nobelprize.org/prizes/medicine/2018/allison/biographical/ (accessed July 9, 2024). See *Jim Allen: Breakthrough*, directed by Bill Haney, Uncommon Productions, 2019, https://www.uncommonproductions.com/breakthrough (accessed July 9, 2024).
27. A chip set refers to the components of integrated circuits that regulate the flow of data in electronic devices such as smartphones.
28. John McCorkle, interview by author, audio recording, Falls Church, VA, December 16, 2022.
29. McCorkle, personal interview.
30. Cephus E. Simmons Sr., interview by author, audio recording, Mt. Pleasant, SC, December 8, 2023.
31. UNC School of Medicine, "Cephus Simmons ('07 MRS) Uses RA Experience and Innovation to Launch New Business," November 22, 2021, https://www.med.unc.edu/healthsciences/radisci/2021/11/cephus-simmons-07-mrs-uses-ra-experience-and-innovation-to-launch-new-business/.
32. Simmons, personal interview.
33. "About Our History," Cade Museum for Creativity and Invention, https://cademuseum.org/about/history/ (accessed June 21, 2024).
34. Phoebe Miles, "To Invent Is Divine: How Strong Patent Rights Help Us to Reflect God's Innovative Nature," with James Edwards, at Faith & Law, *Friday Forum*, July 8, 2022, 2172 Rayburn House Office Building, Washington, DC, video, 12:03, https://faithandlaw.org/resources/2022/05/to-invent-is-divine/.
35. "About Our History," Cade Museum.
36. "About Our History," Cade Museum.
37. Thomas Massie, interview by author, audio recording via telephone, May 14, 2024.
38. Yonover and Crowe, *Hardcore Inventing*, 1.
39. STEM stands for science, technology, engineering, and mathematics.
40. John Tyndall, *Faraday as a Discoverer* (Thomas Crowell, 1961), 23, Google Books, https://books.google.com/books?id=ohNzDzbL_EEC&source=gbs_book_other_versions.
41. See Yonover and Crowe, *Hardcore Inventing*, 48–50.
42. Simmons, personal interview.
43. Simmons, personal interview.
44. Armstrong, personal interview.
45. Armstrong, personal interview.
46. Armstrong, personal interview.
47. Armstrong, personal interview.

Chapter 6: Inventors and Common Grace

1. Michael G. Brown and Zach Keele, *Sacred Bond: Covenant Theology Explored*, 2nd ed. (Reformed Fellowship, 2017), 77.
2. Louis Berkhof, *Systematic Theology*, 2nd ed. (Banner of Truth, 2021), 448.

3. Berkhof, *Systematic Theology*, 444.

4. Michael Molinsky, "Quotations in Context: Galileo – 2," *Convergence*, December 2023, https://old.maa.org/press/periodicals/convergence/quotations-in-context-galileo-2 (accessed June 22, 2024). Molinsky calls this a paraphrase. A prominent English translation to which this quotation is attributed reads, "Philosophy is written in this grand book, the universe, which stands continually open to our gaze. But the book cannot be understood unless one first learns to comprehend the language and read the letters in which it is composed. It is written in the language of mathematics, and its characters are triangles, circles, and other geometric figures without which it is humanly impossible to understand a single word of it; without these, one wanders about in a dark labyrinth." Another rendering has Galileo saying, "The laws of nature are written by the hand of God in the language of mathematics." Paul James-Griffiths, "Quotes from Famous Christians in Science and Medicine," *Christian Heritage Edinburgh*, August 11, 2016, https://www.christianheritageedinburgh.org.uk/category/science/quotes-from-famous-christians-in-science-and-medicine/.

5. Kenneth A. Myers, *All God's Children and Blue Suede Shoes: Christians and Popular Culture* (Crossway, 1989), 35.

6. Derek Kidner, *The Wisdom of Proverbs, Job & Ecclesiastes: An Introduction to Wisdom Literature* (Crossway, 1985), 11, https://books.google.com/books/about/The_Wisdom_of_Proverbs_Job_and_Ecclesias.html?id=u00jCgAAQBAJ.

7. See, e.g., Micah 4:1–5 and Zechariah 3:10.

8. Samuel F. B. Morse, letter to his parents, January 15, 1826, Samuel F. B. Morse Papers, Library of Congress, microfilm reel 25, frame 645.

9. Kenneth Silverman, *Lightning Man: The Accursed Life of Samuel F. B. Morse* (Da Capo, 2004), 82.

10. John McCorkle, interview by author, audio recording, Falls Church, VA, December 16, 2022.

11. Jill Jonnes, *Empires of Light: Edison, Tesla, Westinghouse, and the Race to Electrify the World* (Random House, 2004), 58.

12. Jonnes, *Empires of Light*, 52–53; Joel Martin and William J. Birnes, *Edison vs. Tesla: The Battle Over Their Last Invention* (MJF, 2017), 94–95.

13. "Detailed Biography," Thomas A. Edison Papers, Rutgers Univ. School of Arts and Sciences, October 28, 2016, https://edison.rutgers.edu/life-of-edison/biography/detailed-biography.

14. Adam Mossoff, "The History of Patent Licensing and Secondary Markets in Patents: An Antidote to False Rhetoric," George Mason Univ. Center for Intellectual Property X Innovation Policy, December 9, 2011, https://cip2.gmu.edu/2013/12/09/the-history-of-patent-licensing-and-secondary-markets-in-patents-an-antidote-to-false-rhetoric/.

15. Dudley Nichols, "Electrical Power Straight from the Sun," *Forbes*, June 15, 1929, 20, https://archive.org/details/sim_forbes_1929-06-15_23_12/page/20/mode/1up?view=theater.

16. Jonnes, *Empires of Light*, 352.

17. Kevin Daum, "37 Quotes from Thomas Edison That Will Inspire Success," *Inc.*, February 11, 2016, https://www.inc.com/kevin-daum/37-quotes-from-thomas-edison-that-will-bring-out-your-best.html.

18. Jonnes, *Empires of Light*, 225.
19. Martin and Birnes, *Edison vs. Tesla*, 152.
20. Edward Marshall, "'No Immortality of the Soul' Says Thomas A. Edison," *New York Times Magazine*, October 2, 1910, 1, in the Thomas A. Edison Papers, Rutgers Univ., https://edisondigital.rutgers.edu/document/D1027AAA1 (accessed July 11, 2024).
21. Marshall, "No Immortality of the Soul," 1; Martin and Birnes, *Edison vs. Tesla*, 151–52.
22. Ellen Vaughn, *Time Peace: Living Here and Now with a Timeless God* (Zondervan, 2007), 137.
23. Andy Kessler, "A Faraday Is Worth 1,000 Faucis," Inside View, *Wall Street Journal*, September 19, 2022; Vaughn, *Time Peace*, 136–37.
24. T. R. Reid, *The Chip: How Two Americans Invented the Microchip and Launched a Revolution*, rev. ed. (Random House, 2001), 25.
25. Jonnes, *Empires of Light*, 43.
26. Jonnes, *Empires of Light*, 42.
27. "George Washington Carver: American Agricultural Chemist," *Encyclopaedia Britannica*, https://www.britannica.com/biography/George-Washington-Carver (accessed July 5, 2024).
28. William J. Federer, *George Washington Carver: His Life and Faith in His Own Words* (Amerisearch, 2003), 68.
29. "George Washington Carver," *Encyclopaedia Britannica*.
30. Federer, *George Washington Carver*, 53.
31. Federer, *George Washington Carver*, 54.
32. Federer, *George Washington Carver*, 53.
33. "How George Washington Carver Revolutionized Agriculture," National Inventors Hall of Fame, https://www.invent.org/blog/inventors/george-washington-carver-inventions (accessed July 5, 2024); "George Washington Carver," National Inventors Hall of Fame, https://www.invent.org/inductees/george-washington-carver (accessed July 5, 2024).
34. "George Washington Carver Biography," *Biography*, April 27, 2017, https://www.biography.com/scientist/george-washington-carver.
35. Charles Graeber, "Meet the Carousing, Harmonica-Playing Texan Who Won a Nobel for His Cancer Breakthrough," *Wired*, October 22, 2018, https://www.wired.com/story/meet-jim-allison-the-texan-who-just-won-a-nobel-cancer-breakthrough/.
36. Claudia Dreifus, "The Contrarian Who Cures Cancers," *Quanta Magazine*, February 3, 2020, https://www.quantamagazine.org/the-contrarian-who-cures-cancers-20200203/.
37. See Jo Cavallo, "Immunotherapy Research of James P. Allison, PhD, Has Led to a Paradigm Shift in the Treatment of Cancer," *ASCO Post*, September 15, 2014, https://ascopost.com/issues/september-15-2014/immunotherapy-research-of-james-p-allison-phd-has-led-to-a-paradigm-shift-in-the-treatment-of-cancer/.
38. "Transcript from an Interview with James P. Allison," Nobelprize.org, December 6, 2018, https://www.nobelprize.org/prizes/medicine/2018/allison/159229-james-allison-interview-transcript/.
39. "Transcript from an Interview with James P. Allison," Nobelprize.org.

40. Nuño Domínguez, "Science & Tech Interview with James Allison," *El Pais*, May 20, 2022, https://english.elpais.com/science-tech/2022-05-21/james-allison-the-cost-of-some-cancer-drugs-is-crazy-theres-no-relationship-to-the-cost-of-making-the-drug-anymore.html.

41. Dreifus, "The Contrarian."

42. Colleen Cheslak, "Hedy Lamarr," National Women's History Museum, August 30, 2018, https://www.womenshistory.org/education-resources/biographies/hedy-lamarr.

43. Gerri Miller, "'Bombshell' Delves into the Genius and Jewish Identity of Hedy Lamarr," *Jewish Journal*, May 9, 2018, https://jewishjournal.com/culture/arts/233979/bombshell-delves-genius-jewish-identity-hedy-lamarr/.

44. Cheslak, "Hedy Lamarr."

45. Hannah L. Miller, "How Hedy Lamarr and Her Inventions Changed the World," *Leaders*, August 30, 2022, https://leaders.com/articles/leaders-stories/hedy-lamarr-inventions/.

46. Ruth Barton, "Hedy Lamarr," *Shalvi/Hyman Encyclopedia of Jewish Women*, Jewish Women's Archive, June 23, 2021, https://jwa.org/encyclopedia/article/lamarr-hedy.

47. See Alice George, "Thank This World War II-Era Film Star for Your Wi-Fi," *Smithsonian Magazine* April 4, 2019, https://www.smithsonianmag.com/smithsonian-institution/thank-world-war-ii-era-film-star-your-wi-fi-180971584/, and Miller, "How Hedy Lamarr and Her Inventions Changed the World."

48. Miller, "How Hedy Lamarr and Her Inventions Changed the World."

49. George, "Thank This World War II-Era Film Star for Your Wi-Fi."

50. Thomas Massie, interview by author, audio recording, May 14, 2024.

51. Massie, personal interview.

52. Massie, personal interview.

53. Massie, personal interview.

54. "Raymond Wissolik," *Hendersonville (NC) Times-News*, June 19, 2009, https://www.blueridgenow.com/story/news/2009/06/19/raymond-wissolik/28188260007/.

55. Erica Wissolik, e-mail message to the author, August 18, 2021.

56. "Raymond Damadian," Lemelson-MIT Program, https://lemelson.mit.edu/award-winners/raymond-damadian (accessed July 12, 2024); James R. Hagerty, "Doctor Pioneered MRI Scanning," *Wall Street Journal*, August 27, 2022, A11.

57. Raymond Damadian, "Discovering the MRI Scanner," *Guideposts*, January 1999, 23, quoted in Jerry Bergman, "Raymond Damadian, Inventor of the MRI," Institute for Creation Research, April 30, 2015, https://www.icr.org/article/raymond-damadian-inventor-mri.

Chapter 7: Creativity's By-Products: Human Flourishing

1. Hugh Whelchel, "Six Ways Biblical Flourishing Is Unique," Institute for Faith, Work & Economics, March 3, 2017, https://tifwe.org/six-reasons-biblical-flourishing-is-unique/.

2. Neil G. Messer, "Human Flourishing: A Christian Theological Perspective," in Matthew T. Lee, Laura D. Kubzansky, and Tyler J. VanderWeele, eds., *Measuring Well-Being: Interdisciplinary Perspectives from the Social Sciences and*

the Humanities, online ed. (Oxford Academic, 2021), https://doi.org/10.1093/oso/9780197512531.003.0011, 290.

3. Cornelius Plantinga Jr., *Not the Way It's Supposed to Be: A Breviary of Sin* (Wm. B. Eerdmans, 1995), 37–38.

4. Justin Taylor, "An Interview with Andy Crouch about the Idol and Gift of Power," Gospel Coalition, September 30, 2013, https://www.thegospelcoalition.org/blogs/justin-taylor/an-interview-with-andy-crouch-about-the-idol-and-gift-of-power/.

5. Andy Crouch, *Culture Making: Recovering Our Creative Calling* (InterVarsity, 2008), 107.

6. Arnold Palmer, *A Life Well Played: My Stories* (St. Martin's, 2016), 251.

7. Plantinga, *Not the Way It's Supposed to Be*, 10.

8. John McCorkle, interview by author, audio recording, Falls Church, VA, December 16, 2022.

9. Claire Kendall, interview by author, audio recording, Mt. Pleasant, SC, December 8, 2023.

10. "Phoebe Miles, "To Invent Is Divine: How Strong Patent Rights Help Us to Reflect God's Innovative Nature," with James Edwards, at Faith & Law, *Friday Forum*, July 8, 2022, 2172 Rayburn House Office Building, Washington, DC, video, 9:19, https://faithandlaw.org/resources/2022/05/to-invent-is-divine/. Also, see https://cademuseum.org/about/inventivity-framework/.

11. Edmund S. Phelps, "To Fix Our Economic and Social Malaise, We Need to Rediscover Where Progress Comes From," *MarketWatch*, January 29, 2019, https://www.marketwatch.com/story/to-fix-our-economic-and-social-malaise-we-need-to-rediscover-where-progress-comes-from-2019-01-28.

12. Paul DeRosa, "The Eclectic Brilliance of Edmund Phelps," *American Interest* 9, no. 5, April 20, 2014, https://www.the-american-interest.com/2014/04/20/the-eclectic-brilliance-of-edmund-phelps/.

13. Phelps, "To Fix Our Economic and Social Malaise."

14. Romesh Vaitilingam, "Blog: Innovation and Human Flourishing," *Lindau Nobel Laureate Meetings*, July 25, 2014, https://www.lindau-nobel.org/innovation-and-human-flourishing/.

15. DeRosa, "The Eclectic Brilliance of Edmund Phelps."

16. Crouch, *Culture Making*, 73.

17. Harold Lindsell, *Free Enterprise: A Judeo-Christian Defense* (Tyndale House, 1982), 70.

18. Lindsell, *Free Enterprise*, 71.

19. Madeleine L'Engle, "The Expanding Universe," Newbery Medal acceptance speech, August 1963, printed in *A Wrinkle in Time* (Square Fish, 2007), 244–45.

20. Nancy Pearcey, *Saving Leonardo: A Call to Resist the Secular Assault on Mind, Morals, & Meaning* (B&H, 2010), 83.

21. Whelchel, "Six Ways Biblical Flourishing Is Unique."

22. Jonathan Kandell, "The Glorious History of Handel's Messiah," *Smithsonian Magazine*, December 2009, https://www.smithsonianmag.com/arts-culture/the-glorious-history-of-handels-messiah-148168540/.

23. Taylor, "An Interview with Andy Crouch."

24. See Matthew 15:3–6, where Jesus confronts the Pharisees' regulation that breaks the commandment to "Honor your father and your mother." Instead of their child providing his parents help, the perverse rule claims that "What you would have gained from me is given to God." The Son of God took issue with this manmade rule.

25. Chris Haire, "KIYATEC's Matthew Gevaert Talks about the Greenville Company's Ability to Predict Chemotherapy Success," *Greenville (SC) Business Magazine*, October 10, 2019, http://www.greenvillebusinessmag.com/2019/10/10/287910/kiyatec-s-matthew-gevaert-talks-about-the-greenville-company-s-ability-to-predict-chemotherapy-success; Muriel Vega, "LabCorp-Backed Medtech Startup Predicts Outcome of Cancer Therapies on Tumors," *Hypepopotamus*, June 19, 2019, https://hypepotamus.com/companies/kiyatec/.

26. An example illustrating some of the facets of KIYATEC's flourishing, for its team and investors and for its community, customers, and patients, is found in this press release about one of the company's milestones: "Functional Precision Oncology Leader Kiyatec Announces Series C Round with US$18 Million Closing to Accelerate Adoption of Transformational Cancer Platform," December 12, 2022, https://kiyatec.com/functional-precision-oncology-leader-kiyatec-announces-series-c-round-with-us18-million-closing-to-accelerate-adoption-of-transformational-cancer-platform/.

27. "Bell Laboratories," Britannica.com, updated August 4, 2024, https://www.britannica.com/money/Bell-Laboratories.

Chapter 8: God the Owner

1. Matthew Henry, *A Commentary on the Holy Bible*, illustrated ed., vol. 1 (W. P. Blessing, n.d.), 189.

2. For an in-depth discussion of what the first four commandments entail as required duties and prohibited conduct, see the Westminster Larger Catechism, questions 102–21.

3. Matthew Henry, *A Commentary on the Holy Bible*, vol. 1, 188–89.

Chapter 9: Made in His Image: Human Ownership

1. Harold Lindsell, *Free Enterprise: A Judeo-Christian Defense* (Tyndale House, 1982), 54.

2. Judah and Israel together refer to a united kingdom under David and Solomon; later, these territories divided into separate kingdoms.

3. See Daniel L. Dreisbach, *Reading the Bible with the Founding Fathers* (Oxford Univ., 2017), chap. 10, pp. 211–27.

4. George Washington, "Washington to Marquis de Lafayette, Feb. 1, 1784," *The Writings of George Washington*, vol. 10 (1782–1785), Worthington Chauncey Ford, ed. (G. P. Putnam's Sons, 1891) in the Online Library of Liberty, https://oll.libertyfund.org/titles/ford-the-writings-of-george-washington-vol-x-1782-1785 (accessed July 19, 2024).

5. Matthew Henry, *A Commentary on the Holy Bible*, illustrated ed., vol. 2 (W. P. Blessing, n.d.), 444–45.

6. "Defrauding Our Brothers," *Tabletalk*, March 2021, 38.

7. "Defrauding Our Brothers," *Tabletalk*, 38.

8. Dreisbach, *Reading the Bible with the Founding Fathers*, 220.
9. See Dreisbach, *Reading the Bible with the Founding Fathers*, 221.
10. Dreisbach, *Reading the Bible with the Founding Fathers*, 220.
11. Dreisbach, *Reading the Bible with the Founding Fathers*, 220–21.
12. Dreisbach, *Reading the Bible with the Founding Fathers*, 193.
13. Dreisbach, *Reading the Bible with the Founding Fathers*, 193–94.
14. Lindsell, *Free Enterprise*, 52.
15. Lindsell, *Free Enterprise*, 53.
16. Lindsell, *Free Enterprise*, 53.
17. Lindsell, *Free Enterprise*, 53.

Chapter 10: The Mutual Reinforcement of Creativity and Ownership

1. Abraham Lincoln, "Second Lecture on Discoveries and Inventions," February 11, 1859, *Collected Works of Abraham Lincoln*, vol. 3, August 21, 1858–March 4, 1860, Univ. of Michigan Library Digital Collections, https://name.umdl.umich.edu/lincoln3 (accessed July 20, 2024), 363.

2. Lincoln, "Second Lecture on Discoveries and Inventions," 363.

3. Plato, *Republic*, Benjamin Jowett, trans. (Barnes & Noble, 2004), 347. "Necessity Is the Mother of Invention," Phrases.org.uk, Gary Martin, https://www.phrases.org.uk/meanings/necessity-is-the-mother-of-invention.html (accessed July 13, 2024).

4. Mary Bellis, "The Invention of the Wheel," ThoughtCo, updated July 14, 2024, https://www.thoughtco.com/the-invention-of-the-wheel-1992669.

5. Jill Jonnes, *Empires of Light: Edison, Tesla, Westinghouse, and the Race to Electrify the World* (Random House, 2004), 62.

6. Jonnes, *Empires of Light*, 58–65.

7. Dudley Nichols, "Electrical Power Straight from the Sun," *Forbes*, June 15, 1929, 20, https://archive.org/details/sim_forbes_1929-06-15_23_12/page/20/mode/1up?view=theater.

8. Charles D. Lanier, "Two Giants of the Electrical Age," *Review of Reviews*, vol. 8 (1893): 41.

9. "Magnificent Power Celebration Banquet at the Ellicott Club," *Buffalo (NY) Morning Express*, January 13, 1897, 1. Tesla's "On Electricity" speech also appeared printed in the *Electrical Review*, January 27, 1897, with slight variation from the Buffalo newspaper account. This could be due to the reporter's inadvertent changes in wording or Tesla's polishing his text for posterity. The latter version is available several places online, including at PBS, https://www.pbs.org/tesla/res/res_art04.html.

10. William J. Federer, *George Washington Carver: His Life and Faith in His Own Words* (Amerisearch, 2003), 53.

11. Federer, *George Washington Carver*, 68.

12. Susan M. Armstrong, interview by author, audio recording via Skype, October 23, 2023.

Chapter 11: Benefits from Owning What You Create

1. Thomas Massie, interview by author, audio recording via telephone, May 14, 2024.

2. Elizabeth Elizalde, "Famed Cellist Yo-Yo Ma Turns MA Vaccination Site into Concert Hall," *New York Post,* March 14, 2021, https://nypost.com/2021/03/14/yo-yo-ma-turns-ma-vaccination-site-into-concert-hall/.

3. "U.S. Patent Issued for Three-Point Seatbelt," This Day In History: July 10, history.com, https://www.history.com/this-day-in-history/u-s-patent-issued-for-three-point-seatbelt (accessed July 20, 2024).

4. See Francis S. Collins, *The Language of God: A Scientist Presents Evidence for Belief* (Free Press, 2006), 117–22.

5. SCbio, "Study Shows Typical Cancer-Free Survival Doubled for Recurrent Brain Cancer Patients When KIYATEC's Test Informed Therapy Selection," December 18, 2020, https://www.scbio.org/articles/study-shows-typical-cancer-free-survival-doubled-for-recurrent-brain-cancer-patients-when-kiyatecs-test-informed-therapy-selection.

6. Chris Haire, "KIYATEC's Matthew Gevaert Talks about the Greenville Company's Ability to Predict Chemotherapy Success," *Greenville (SC) Business Magazine,* October 10, 2019, http://www.greenvillebusinessmag.com/2019/10/10/287910/kiyatec-s-matthew-gevaert-talks-about-the-greenville-company-s-ability-to-predict-chemotherapy-success.

7. SCbio, "KIYATEC Announces Investment from Seae Ventures and Names Managing Partner Jason Robart to KIYATEC Board of Directors," May 5, 2021, https://www.scbio.org/articles/kiyatec-announces-investment-from-seae-ventures-and-names-managing-partner-jason-robart-to-kiyatec-board-of-directors; Muriel Vega, "LabCorp-Backed Medtech Startup Predicts Outcome of Cancer Therapies on Tumors," *Hypepopotamus,* June 19, 2019, https://hypepotamus.com/companies/kiyatec/.

8. KIYATEC news release, "Kiyatec Announces Investment from Leading Brain Cancer Venture Philanthropy Funds," May 18, 2023, https://www.kiyatec.com/kiyatec-announces-investment-from-leading-brain-cancer-venture-philanthropy-funds.

9. See "Connectivity Changes Everything," 2018–2020 Qualcomm Exhibit, National Inventors Hall of Fame; and "Connecting Everything Everywhere: Public Policy to Ensure United States Leadership in 5G," Qualcomm, 2019. Relevant here, Qualcomm owns mobile GPS U.S. patent no. 7,876,265 and app-sharing technology patent no. 7,099,663.

10. A "value chain" is the sequence of "links" spanning from raw materials to other inputs constituting an end product. Each link in the chain adds a certain amount of value to an eventual product. Foundational technology provides an outsized level of value to an end product due to the innovation it embodies. Foundational technological components, for example, make possible certain functions, provide greater capacity, faster speeds, etc., while less valuable elements in a value chain serve more marginal purposes. For more information, see "The Value Chain," Harvard Business School Institute for Strategy & Competitiveness, https://www.isc.hbs.edu/strategy/business-strategy/Pages/the-value-chain.aspx (accessed August 9, 2024).

11. Justin Taylor, "An Interview with Andy Crouch about the Idol and Gift of Power," Gospel Coalition, September 30, 2013, https://www.thegospelcoalition.org/blogs/justin-taylor/an-interview-with-andy-crouch-about-the-idol-and-gift-of-power/.

12. Cephus Simmons, interview by author, audio recording, Mt. Pleasant, SC, December 8, 2023.

13. *Chariots of Fire*, directed by Hugh Hudson, Warner Bros. Pictures, 1981.

Chapter 12:

1. David S. Landes, *The Wealth and Poverty of Nations: Why Some Are So Rich and Some So Poor* (W.W. Norton, 1998), 31–32.

2. Landes, *The Wealth and Poverty of Nations*, 217–18.

3. Westminster Larger Catechism in R. C. Sproul, ed., *Reformation Study Bible* (Ligonier Ministries, 2008), 2464.

4. "The Fifth Commandment," Ligonier Ministries, October 11, 2012, https://www.ligonier.org/learn/devotionals/the-fifth-commandment. See, for instance, Romans 13:1–7 regarding obedience to civil authorities.

5. "The Seventh Commandment," Ligonier Ministries, September 6, 2010, https://www.ligonier.org/learn/devotionals/the-fifth-commandment.

6. "The Ninth Commandment," Ligonier Ministries, November 6, 2012, https://www.ligonier.org/learn/devotionals/the-ninth-commandment.

7. John Piper, "How Should Christians Think About Socialism?", October 20, 2015, https://www.desiringgod.org/interviews/how-should-christians-think-about-socialism.

8. Harold Lindsell, *Free Enterprise: A Judeo-Christian Defense* (Tyndale House, 1982), 42.

9. Lindsell, *Free Enterprise*, 14.

10. Lindsell, *Free Enterprise*, 15.

11. John Locke's articulation of natural rights, consent-based government, property rights, etc., may have popularized these concepts, including with many American Founding Fathers, but the Reformation deserves credit for its associated benefits from returning the Bible to an authority, including over monarchs. Christian thinker Francis A. Schaeffer emphasizes that the Protestant Reformers' preaching the Gospel and teaching the Bible sparked reforms that affected not only religion, but law, government, and other societal spheres. Schaeffer notes in his book *How Should We Then Live?* (chap. 5) how Locke drew deeply from Scottish Reformer Samuel Rutherford's 1644 *Lex Rex* (Law Is King), which set forth arguments and reasoning from Scripture for government under law rather than by arbitrary monarchal rule. Schaeffer says Locke advanced *Lex Rex*'s arguments without their biblical underpinnings. Notably, another source in addition to Locke influenced the American Founders with principles of *Lex Rex*, Schaeffer says. John Witherspoon, a Scottish Presbyterian minister and devotee of Rutherford's *Lex Rex*, became president of the College of New Jersey (now Princeton) before the Revolutionary War. Witherspoon served as a delegate from New Jersey to the Continental Congress and signed the Declaration of Independence.

12. "Lockean Labor Theory Law and Legal Definition," USLegal, https://definitions.uslegal.com/l/lockean-labor-theory/ (accessed April 8, 2023).

13. John Locke, *Two Treatises of Government*, 2, sec. 27, in the Online Library of Liberty, https://oll.libertyfund.org/title/hollis-the-two-treatises-of-civil-government-hollis-ed#lf0057_label_207 (accessed December 29, 2023).

14. Locke, *Second Treatise of Government*, sec. 44.

15. Locke, *Second Treatise of Government*, sec. 27.

16. Locke, *Second Treatise of Government*, sec. 37.

17. For more on Lockean property rights and how labor serves to create wealth, see Edward J. Erler, *Property and the Pursuit of Happiness: Locke, the Declaration of Independence, Madison, and the Challenge of the Administrative State* (Rowman & Littlefield, 2019) and Thomas G. West, *Vindicating the Founders: Race, Sex, Class, and Justice in the Origins of America* (Rowman & Littlefield, 1997), chap. 2, "Property Rights."

18. Locke, *Second Treatise of Government*, sec. 32.

19. The Renaissance, roughly from the 1300s to the 1600s, was a "conscious rebirth of ancient culture" that renewed interest in classical Greek and Roman arts and literature, architecture, and other forms of "humanism." The Renaissance produced much discovery, art, literature, architecture, and other creative works, though generally it raised humanity to the "divine" and lowered God. The Protestant Reformation influenced the American founding far more than did the Renaissance—"Early America had neither the wealth nor the leisure for splendid cultivation of the arts or of letters. . . . Rather, colonial America generally shared the Reformers' detestation of Renaissance notions and ways." Russell Kirk, *The Roots of American Order*, 3rd ed. (Regnery Gateway, 1991), 223, 228, 230–31, 235.

20. Samuel P. Huntington, *The Clash of Civilizations and the Remaking of World Order* (Simon & Schuster, 1996), 46–55, 68–72. Commonwealth refers to the mutual best interests and rights of the entire citizenry. *Black's Law Dictionary* calls it a republican form of government focused on the general welfare rather than the privileges of a few. A republic is a government where representatives chosen by a nation's citizens conduct government on their behalf (Henry Campbell Black, "Republic," "Republican Government," *Black's Law Dictionary*, 6th ed. (West Publishing, 1990), 1302, 1303. Russell Kirk's *The Roots of American Order* 3rd ed. says a republic "has no hereditary monarch" and "may be either aristocratic or democratic" (Regnery, 1991), 415). Democracy is government by the whole citizenry, exercised directly or through representatives.

21. Declaration of Independence, National Archives, https://www.archives.gov/founding-docs/declaration (accessed December 30, 2023).

22. Erler, *Property and the Pursuit of Happiness*, 3.

23. Virginia Declaration of Rights, National Archives, https://www.archives.gov/founding-docs/virginia-declaration-of-rights (accessed December 30, 2023).

24. U.S. Constitution, National Archives, https://www.archives.gov/founding-docs/constitution (accessed December 30, 2023).

25. James Madison, Federalist 10, in Benjamin F. Wright, ed., *The Federalist* (Barnes & Noble Books, 2004), 130, 131.

26. Locke, *Second Treatise of Government*, sec. 124.

27. James Madison, "Property," *The Writings*, vol. 6, 1790–1802 (G. P. Putnam's Sons, 1906), in the Online Library of Liberty, https://oll.libertyfund.org/titles/madison-the-writings-vol-6-1790-1802 (accessed July 23, 2024).

28. Kirk's *The Roots of American Order* describes the democratic republican government found in the Constitution: "The American Republic was a government on a national scale, but with the powers of the general government limited; also the powers

of that Republic's component states were restricted. The most remarkable features of this Republic would be its independent national judiciary, endowed with power to rule upon the constitutionality of the acts of national and state legislatures; and its successful 'federal' character, 'out of many, one,' reconciling national needs and self-government in its member states. It would be a democracy of elevation, not of mediocrity, with strong guarantees for the security of life, liberty, property, and other private rights" (415).

29. John Adams, *The Works of John Adams*, vol. 6 (Little, Brown, 1851), in the Online Library of Liberty, https://oll.libertyfund.org/title/adams-the-works-of-john-adams-vol-6 (accessed December 28, 2023).

30. Huntington, *The Clash of Civilizations*, 46.

31. Edmund Phelps, *Mass Flourishing: How Grassroots Innovation Created Jobs, Challenge, and Change* (Princeton Univ., 2013), 177.

32. Phelps, *Mass Flourishing*, 177.

33. Phelps, *Mass Flourishing*, 178.

34. Matthew Spalding, *We Still Hold These Truths: Rediscovering Our Principles, Reclaiming Our Future* (ISI, 2009), 74–75.

35. Ronald Reagan, Speech on Project Economic Justice, the White House, Washington, DC, August 3, 1987, https://www.cesj.org/about-cesj-in-brief/history-accomplishments/pres-reagans-speech-on-project-economic-justice/.

36. Russell Kirk, *The Politics of Prudence* (ISI, 1993), 20.

Chapter 13

1. Robert Yonover and Ellie Crowe, *Hardcore Inventing: Invent, Protect, Promote, and Profit from Your Inventions* (New York: Skyhorse, 2009), 89.

2. *Davoll v. Brown*, 7 F. Cas. 197 (C.C.D. Mass., 1845).

3. Pat Choate, *Hot Property: The Stealing of Ideas in an Age of Globalization* (Alfred A. Knopf, 2005), 13.

4. Choate, *Hot Property*, 13.

5. The term "equity" here means one's share of ownership in certain property, in this case an invention or creative work. This type of equity is akin to a shareholder's proportionate holdings in a company's capital stock or the amount of equity a homeowner holds in his or her home—how much of the principal amount of a mortgage loan plus the amount of the purchase price paid in a down payment when the house or condominium purchased has been paid off. In the latter instance, banks make home equity loans to homeowners, a second mortgage based on how much equity, or house, the owners have paid off and thus own outright.

6. Yonover and Crowe, *Hardcore Inventing*, 89.

7. Henry Campbell Black, "Intangible property," *Black's Law Dictionary*, 6th ed. (West Publishing, 1990), 809.

8. Siskind's illustration of IP protection appears in Choate, *Hot Property*, 14–15.

9. Kent R. Middleton and Bill F. Chamberlin, *The Law of Public Communication*, 3rd ed. (Longman, 1994), 214.

10. Choate, *Hot Property*, 26–27.

11. See Middleton and Chamberlin, *The Law of Public Communication*, 214ff.

12. "Copyright," *Black's Law Dictionary*, 336.

13. Choate, *Hot Property*, 261.

14. "Fair use doctrine," *Black's Law Dictionary*, 598.

15. See *Black's Law Dictionary*, 598–99; Middleton and Chamberlin, *The Law of Public Communication*, 231–41; and Choate, *Hot Property*, 261–62.

16. "FAQ's: How Long Does Copyright Protection Last?", U.S. Copyright Office, https://www.copyright.gov/help/faq/faq-duration.html (accessed February 3, 2024).

17. See *Black's Law Dictionary*, 1606, and Middleton and Chamberlin, *The Law of Public Communication*, 221–22.

18. Francina Cantatore and Elizabeth Crawford-Spencer, *Effective Intellectual Property Management for Small to Medium Businesses and Social Enterprises* (Brown-Walker, 2018), 3.

19. *Black's Law Dictionary*, "Patent," 1125.

20. Robert M. Weir, *Colonial South Carolina: A History* (Univ. of South Carolina, 1997), 49–53.

21. *Black's Law Dictionary*, "Utility patent," 1125.

22. *Black's Law Dictionary*, "Design patent," 1125.

23. *Black's Law Dictionary*, "Plant patent," 1125.

24. *Black's Law Dictionary*, 1125. Also, see Title 35 of the U.S. Code.

25. Choate, *Hot Property*, 295.

26. *Black's Law Dictionary*, "Patent-right," 1125.

27. Cantatore and Crawford-Spencer, *Effective Intellectual Property Management*, 3.

28. Choate, *Hot Property*, 199.

29. Yonover and Crowe, *Hardcore Inventing*, 104.

30. Charles B. McGough, *Great Invention! Now What?: Evaluate, Patent, Trademark, and License Your New Invention* (Self-Counsel, 2014), 28. Also, see Yonover and Crowe, *Hardcore Inventing*, 104.

31. To clarify, ideas as such are not eligible for IP protection. Ideas must be realized in a tangible form, embodied in a medium, applied in an article, or otherwise have undergone human intervention.

32. Harold Lindsell, *Free Enterprise: A Judeo-Christian Defense* (Tyndale House, 1982), 53.

33. *Black's Law Dictionary*, 1493.

34. Choate, *Hot Property*, 296.

35. See Cantatore and Crawford-Spencer, *Effective Intellectual Property Management*, 3, 76–77; and Choate, *Hot Property*, 296.

36. Choate, *Hot Property*, 39.

37. *Black's Law Dictionary*, "Trade secret," 1494.

38. Choate, *Hot Property*, 296.

39. "Know-how," *Webster's Encyclopedic Unabridged Dictionary of the English Language* (Gramercy, 1996), 793.

40. Choate, *Hot Property*, 296.

41. See Cantatore and Crawford-Spencer, *Effective Intellectual Property Management*, 102ff.

42. Choate, *Hot Property*, 199.

43. Christopher T. Zirpoli, "An Introduction to Trade Secrets Law in the United States," *In Focus* IF12315, Congressional Research Service, January 27, 2023, https://crsreports.congress.gov/product/pdf/IF/IF12315.

44. *Ruckelshaus v. Monsanto*, 467 U.S. 986 (1984).

45. Adam Mossoff, "The Constitutional Protection of Intellectual Property," Heritage Foundation Report, March 8, 2021, https://www.heritage.org/economic-and-property-rights/report/the-constitutional-protection-intellectual-property.

46. Choate, *Hot Property*, 296.

Chapter 14: The American Patent System and the Iconic Inventor

1. Charles Wilbanks, ed., *The American Revolution and Righteous Community: Selected Sermons of Bishop Robert Smith* (Univ. of South Carolina, 2007), 179.

2. "Revolutionary War," History.com, https://www.history.com/topics/american-revolution/american-revolution-history (accessed February 13, 2024).

3. See M. E. Bradford, *A Worthy Company: Brief Lives of the Framers of the United States Constitution* (Plymouth Rock Foundation, 1982), and B. J. Lossing, *Lives of the Signers of the Declaration of Independence*, repr. of 1848 original (Wallbuilders, 2007).

4. See Lossing, *Lives of the Signers*, 180.

5. Pat Choate, *Hot Property: The Stealing of Ideas in an Age of Globalization* (Alfred A. Knopf, 2005), 24.

6. John W. Oliver, *History of American Technology* (Ronald, 1956), 89.

7. Choate, *Hot Property*, 24.

8. Though outside the scope of this book, an important initiative for overcoming the United States' economic and security vulnerabilities bears mention. Treasury Secretary Alexander Hamilton's "Report on Manufactures," given to Congress in 1791, lays out the case for making domestic manufacturing a key component of the U.S. economy. Hamilton sought to stimulate an industrial sector capable of making military and other goods, to diversify the U.S. economy, to create wealth, to make better use of America's natural and human resources, and to provide sound footing for American society and secure independence. See, for instance, Richard Brookhiser, *Alexander Hamilton: American* (Free Press, 1999), 92–97. The connection of Hamilton's vision for domestic manufacturing to patents and invention comes at commercialization—providing a vehicle for taking an invention to practical usage as a product or other commercially salable good or service.

9. See Bradford, *A Worthy Company*, 66–75; Lossing, *Lives of the Signers*, 104–11.

10. See Bradford, *A Worthy Company*, 68–69; Lossing, *Lives of the Signers*, 68–69; "Benjamin Franklin's Inventions," Franklin Institute, https://fi.edu/en/science-and-education/benjamin-franklin/inventions (accessed February 16, 2024); and "Benjamin Franklin Biography," Biography.com, A&E Television Networks, last updated December 10, 2020, https://www.biography.com/political-figures/benjamin-franklin.

11. "Moldboard Plow," *Thomas Jefferson Encyclopedia*, https://www.monticello.org/research-education/thomas-jefferson-encyclopedia/moldboard-plow/ (accessed February 16, 2024).

12. "Wheel Cipher," *Thomas Jefferson Encyclopedia,* https://www.monticello.org/research-education/thomas-jefferson-encyclopedia/wheel-cipher/ (accessed February 16, 2024).

13. "Roofing," *Thomas Jefferson Encyclopedia,* https://www.monticello.org/research-education/thomas-jefferson-encyclopedia/roofing/ (accessed February 16, 2024).

14. "Patents," *Thomas Jefferson Encyclopedia,* https://www.monticello.org/research-education/thomas-jefferson-encyclopedia/patents/#fn-1 (accessed February 16, 2024).

15. "Patents," *Thomas Jefferson Encyclopedia.*

16. Edward J. Erler, *Property and the Pursuit of Happiness: Locke, the Declaration of Independence, Madison, and the Challenge of the Administrative State* (Rowman & Littlefield, 2019), 93.

17. Erler, *Property and the Pursuit of Happiness,* 94.

18. Erler, *Property and the Pursuit of Happiness,* 94, 95.

19. Choate, *Hot Property,* 27.

20. "The Debates in the Federal Convention of 1787, Which Framed the Constitution of the United States of America," reported by James Madison, September 5, Avalon Project: Madison Debates, Yale Law School, Lillian Goldman Law Library, https://avalon.law.yale.edu/18th_century/debates_905.asp (accessed February 16, 2024); Choate, *Hot Property,* 27; B. Zorina Khan, "History Matters: National Innovation Systems and Innovation Policies in Nations," in Stephen H. Haber and Naomi R. Lamoreaux, eds., *The Battle over Patents: History and Politics of Innovation* (Oxford Univ., 2021), 341. Phyllis Schlafly, who served on the Commission on the Bicentennial of the U.S. Constitution, notes that the unanimous vote to adopt the IP clause was preceded just days before by convention delegates observing steamboat experiments on the Delaware River. Ed Martin, ed., *Phyllis Schlafly Speaks,* vol. 4, Patents & Invention (Skellig America, 2018), 18.

21. Adam Mossoff, "The Constitutional Protection of Intellectual Property," Heritage Foundation, March 8, 2021, https://www.heritage.org/economic-and-property-rights/report/the-constitutional-protection-intellectual-property.

22. Choate, *Hot Property,* 26–27.

23. Mossoff, "The Constitutional Protection of Intellectual Property." In the omitted footnote, Mossoff references his work, "Institutional Design in Patent Law: Private Property Rights or Regulatory Entitlements," 92 *So. Cal. L. Rev.* 921, 936–37 (2019), 921–22. Khan seems to agree; see Khan, "History Matters," 341–42.

24. Choate, *Hot Property,* 27. Khan refers to the resulting U.S. innovation system as "unique in its objective and structure relative to any other in the world," "History Matters," 341.

25. Richard Brookhiser, *Alexander Hamilton: American* (Free Press, 1999), 75.

26. Choate, *Hot Property,* 27.

27. Benjamin F. Wright, ed., *The Federalist* (Barnes & Noble, 2004), 309.

28. Mossoff, "The Constitutional Protection of Intellectual Property."

29. *Evans v. Jordan & Morehead,* 13 U.S. (9 Cranch) 199 (1815), quoted in Mossoff, "The Constitutional Protection of Intellectual Property." The Supreme Court later rejected this in *Wheaton v. Peters,* 33 U.S. (8 Pet.) 591 (1834), holding that the Copyright and Patent Clause does not secure a preexisting right either in copyright or patents. Mossoff helpfully explains in "Who Cares What Thomas Jefferson Thought

About Patents?", *Cornell Law Review* 92 (2007): 985–89, https://papers.ssrn.com/sol3/papers.cfm?abstract_id=892062 how the conventional interpretation of the basis of the *Wheaton* court's holding errs because of misunderstood application of natural rights principles.

30. For a more complete treatment of America's unique democratization of patents, see B. Zorina Khan, *The Democratization of Invention: Patents and Copyrights in American Economic Development* (Cambridge Univ., 2005).

31. Phoebe Miles, "To Invent Is Divine: How Strong Patent Rights Help Us to Reflect God's Innovative Nature," with James Edwards at Faith & Law, Friday Forum, July 8, 2022, 2172 Rayburn House Office Building, Washington, DC, video, 12:02, https://faithandlaw.org/resources/2022/05/to-invent-is-divine/.

32. Patent Act of 1790, sec. 2, Univ. of New Hampshire, https://ipmall.law.unh.edu/sites/default/files/hosted_resources/lipa/patents/Patent_Act_of_1790.pdf (accessed July 26, 2024).

33. Procedurally, the 1790 Patent Act was restrictive. Cabinet members—Secretary of State Thomas Jefferson, Secretary of War Henry Knox, and Attorney General Edmund Randolph—comprised the patent examination panel, and President Washington signed each issued patent. This was a time-consuming task, resulting in a snail's pace of patent review. The 1793 Patent Act removed the dam that blocked patent approvals. Examination was cast aside. Inventors could simply register their inventions with the State Department, and courts would settle patent disputes. The 1836 Patent Act struck a balance and bolstered patent rights. It created a dedicated Patent Office with an extensive patent library, provided examination—not by Cabinet members—to ensure the claimed invention was in fact new, and awarded a patent on competing inventions to the first person to invent the novel invention. Patent applicants were extended the right to appeal denial of a patent to impartial examiners. Patents carried terms of fourteen years of exclusivity from the date of patent issuance. Clear title for the original inventor to the invention and exclusivity for a time period gave patent owners certainty regarding their IP. The key 1836 elements, characterized by awarding patents on the basis of merit and objectivity, remained intact for more than a century and a half. See Choate, *Hot Property*, 27–29, 56.

34. William J. Bennett, *America: The Last Best Hope; From the Age of Discovery to a World at War, 1492–1914*, vol. 1 (Thomas Nelson, 2006), 50.

35. Clarence B. Carson, *A Basic History of the United States: The Colonial Experience, 1607–1774*, vol. 1 (American Textbook Committee, 1990), 98.

36. DeWitte T. Holland, *The Preaching Tradition: A Brief History* (Abingdon, 1980), 55–56.

37. Arnold Dallimore, *George Whitefield: The Life and Times of the Great Evangelist of the Eighteenth Century Revival*, vol. 1 (Banner of Truth, 1971), 413.

38. Dallimore, *George Whitefield*, 413.

39. Daniel L. Dreisbach, *Reading the Bible with the Founding Fathers* (Oxford Univ., 2017), 49.

40. Dreisbach, *Reading the Bible with the Founding Fathers*, 50.

41. Daniel J. Boorstin, *The Americans: The Colonial Experience* (Phoenix, 2000), 323.

42. Carson, *A Basic History of the United States*, vol. 1, 129–30; Boorstin, *The Americans: The Colonial Experience*, 323.

43. Dallimore, *George Whitefield*, 439.

44. Dreisbach, *Reading the Bible with the Founding Fathers*, 51–52.

45. Dreisbach, *Reading the Bible with the Founding Fathers*, 66. See also pages 2–3.

46. Dreisbach, *Reading the Bible with the Founding Fathers*, 66.

47. Dreisbach, *Reading the Bible with the Founding Fathers*, 49.

48. Rod Andrew Jr., *The Life and Times of General Andrew Pickens: Revolutionary War Hero, American Founder* (Univ. of North Carolina, 2017), x.

49. Holland, *The Preaching Tradition*, 55–60; Carson, *A Basic History of the United States*, vol. 1, 100.

50. Reformed faith refers to tenets of belief arising from the Protestant Reformation. Leaders associated with exposition of Reformed theology include Martin Luther, John Calvin, and Ulrich Zwingli. Reformed faith is summarized by grace alone, faith alone, Christ alone, Scripture alone, and God's glory alone. It says God's doctrines of grace satisfy God's perfect justice, the only way to escape the consequences of sin being Jesus's sacrifice of himself on the cross and God's revealing himself through the Bible. Related frameworks are Covenant theology and the five points of Calvinism. See Jonathan Master, "What Is Reformed Theology?" Ligonier Ministries, August 18, 2023, https://www.ligonier.org/learn/articles/what-is-reformed-theology.

51. Dallimore, *George Whitefield*, 587.

52. Dallimore, *George Whitefield*, 588–89.

53. Carson, *A Basic History of the United States*, vol. 1, 100.

54. Bennett, *America*, 52.

55. Holland, *The Preaching Tradition*, 60.

56. Carson, *A Basic History of the United States*, vol. 1, 100.

57. The fuller quotation of Galileo Galilei is, "Mathematics is the language in which God has written the universe. The laws of nature are written by the Hand of God in the language of mathematics. God is known by nature in His works, and by doctrine in His revealed Word." This appears in Galileo's 1623 work, *The Assayer*, https://www.loc.gov/item/2021666740/ (accessed July 27, 2024). Countless translations of this quote appear in academic papers, popular articles, and on websites that curate famous quotations. See Bill Federer, "Early Astronomers: 'Mathematics Is the Language in Which God Has Written the Universe,'" *World Tribune*, August 25, 2018, https://worldtribune.com/life/early-astronomers-mathematics-is-the-language-in-which-god-has-written-the-universe/#google_vignette.

58. Wright, *The Federalist*, 142.

59. Alexis de Tocqueville, *Democracy in America* (Penguin, 2003), 526.

60. The central importance of religious beliefs and practices is exemplified in the example of the Scotch-Irish, such as the extended family of Andrew Pickens. In their settlements from Pennsylvania to Virginia to the Carolinas, the devout Presbyterians building homes on grounds granted by colonial land patents also gathered to hear visiting preachers in open-air locations. They soon built a meeting house for their community's church congregation and practiced familial piety. The Carolinas' state-line Waxhaws community, to where the Pickenses moved around 1753, "had

a permanent minister at least by 1756" despite a shortage of trained ministers, and "weekly church attendance was high" (p. 10). See Rod Andrew Jr., *The Life and Times of General Andrew Pickens*, x–xviii, 3–12.

61. Landes, *The Wealth and Poverty of Nations*, 297.

62. Nathan Rosenberg, ed., *The American System of Manufactures: The Report of the Committee on the Machinery of the United States 1855 and the Special Reports of George Wallis and Joseph Whitworth* (Edinburgh Univ., 1969), 204.

63. Kenneth Silverman, *Lightning Man: The Accursed Life of Samuel F. B. Morse* (Da Capo, 2004), 212, 236.

64. See Schlafly, *Patents and Invention*, 21, 24–25.

65. Charlotte Montague, *Women of Invention: Life-Changing Ideas by Remarkable Women* (Chartwell, 2018), 36, 31.

66. See Hon. Ryan T. Holte, "Clarity in Patent Remedies," *George Mason Law Review* 26, no. 1 (2018): 144–57, https://papers.ssrn.com/sol3/papers.cfm?abstract_id=3492417; and Montague, *Women of Invention*, 120.

67. Choate, *Hot Property*, 65.

68. Mossoff, "The Constitutional Protection of Intellectual Property."

69. Margaret E. Hirst, *The Life of Friedrich List and Selections from His Writings* (Smith, Elder, 1909), 35.

70. Quoted in Kat Eschner, "How Mark Twain's Hatred of Suspenders Drove Him to Invent," *Smithsonian Magazine*, December 19, 2017, https://www.smithsonianmag.com/smart-news/how-mark-twains-hatred-suspenders-drove-him-invent-180967577/.

71. J. J. Harbster and Nate Smith, "The Peripatetic U. S. Patent Office: Locations 1790 to Present," *Inside Adams*, Library of Congress blog ISSN 2691-3690, July 13, 2020, https://blogs.loc.gov/inside_adams/2020/07/the-patent-office/.

72. Silverman, *Lightning Man*, 422; Christian Schussele, *Men of Progress*, 1862, National Portrait Gallery, https://www.si.edu/object/men-progress:npg_NPG.65.60 (accessed March 2, 2024).

73. Dennis Crouch, "Tracing the Quote: Everything That Can Be Invented Has Been Invented," *Patentlyo*, January 6, 2011, https://patentlyo.com/patent/2011/01/tracing-the-quote-everything-that-can-be-invented-has-been-invented.html.

74. These figures are of utility patents only; they do not include design patents, which were first issued in 1843, or plant patents, which began to be granted in 1931.

75. "Milestones in U.S. Patenting," U.S. Patent and Trademark Office, https://www.uspto.gov/patents/milestones (accessed March 2, 2024).

Chapter 15: The Modern U.S. Patent System

1. Pat Choate, *Hot Property: The Stealing of Ideas in an Age of Globalization* (Alfred A. Knopf, 2005), 45.

2. Fred Howard, *Wilbur and Orville: A Biography of the Wright Brothers* (Dover Publications, 1998), 362.

3. Dennis Crouch, "Tracing the Quote: Everything That Can Be Invented Has Been Invented," *Patentlyo*, January 6, 2011, https://patentlyo.com/patent/2011/01/tracing-the-quote-everything-that-can-be-invented-has-been-invented.html.

4. "Milestones in U.S. patenting," U.S. Patent and Trademark Office (USPTO), https://www.uspto.gov/patents/milestones, (accessed March 2, 2024).

5. Stephen H. Haber and Naomi R. Lamoreaux, eds., "Introduction: The Battle over the Surplus from Innovation," in *The Battle over Patents: History and Politics of Innovation* (Oxford Univ., 2021), 22–23; Choate, *Hot Property*, 27–29.

6. Haber and Lamoreaux, eds., "Introduction," *The Battle over Patents*, 22–23.

7. Choate, *Hot Property*, 29, 47, 56; also see Christopher Beauchamp, "Dousing the Fires of Patent Litigation," in Haber and Lamoreaux, eds., *The Battle over Patents*, 146; B. Zorina Khan, "History Matters: National Innovation Systems and Innovation Policies in Nations," in Haber and Lamoreaux, eds., *The Battle over Patents*, 347–49.

8. "Milestones in U.S. patenting," USPTO.

9. Khan, "History Matters," *The Battle over Patents*, 347. For a more complete treatment of America's unique democratization of patents, see B. Zorina Khan, *The Democratization of Invention: Patents and Copyrights in American Economic Development* (Cambridge Univ., 2005).

10. B. Zorina Khan and Kenneth L. Sokoloff, "Lives of Invention: Patenting and Productivity Among Great Inventors in the United States (1790–1930)," 181–99, in Marie-Sophie Corcy, Christiane Douyère-Demeulenaere, Liliane Hilaire-Pérez, eds., *Les Archives de l'invention* (Presses Universitaires du Midi, 2006), 194, https://research.bowdoin.edu/zorina-khan/files/2021/07/Lives-of-Invention.pdf.

11. Robert Yonover and Ellie Crowe, *Hardcore Inventing: Invent, Protect, Promote, and Profit from Your Inventions* (Skyhorse, 2009), 97.

12. Gene Quinn, "Patentability: The Adequate Description Requirement of 35 U.S.C. 112," *IPWatchdog*, June 24, 2017, https://ipwatchdog.com/2017/06/24/patentability-adequate-description-requirement-35-u-s-c-112/id=85039/.

13. Yonover and Crowe, 99–100.

14. Gene Quinn, "How to Write a Patent Application," *IPWatchdog*, August 11, 2018, https://ipwatchdog.com/2018/08/11/how-to-write-a-patent-application/id=97223/.

15. Beauchamp, "Dousing the Fires of Patent Litigation," 150.

16. Beauchamp, "Dousing the Fires of Patent Litigation," 150.

17. The Intellectual Property Owners Association (IPO) describes the grace period: "§102(b) provides that an invention is not prior art if it was patented or described in printed publication anywhere, or in public use or on sale in the U.S., one year or less before the U.S. filing date. 'On sale' is defined in case law." "Comparison of Selected Sections of Pre-AIA and AIA U.S. Patent Law," 1st ed., IPO, October 19, 2011, https://ipo.org/wp-content/uploads/2013/03/Patent_Reform_Chart_Comparison_of_AIA_and_Pre-AIA_Laws_FINAL.pdf.

18. James Edwards, "Legislative Steps in the Pro-Patent Direction," *IPWatchdog*, July 8, 2018, https://ipwatchdog.com/2018/07/08/legislative-steps-pro-patent-direction/id=99068/.

19. "Guide to Intellectual Property: What Is a Patent Model?" National Inventors Hall of Fame *Intellectual Property* (blog), January 29, 2020, https://www.invent.org/blog/intellectual-property/patent-model.

20. Yvonne Morris, "It's All in the Hardware: Overcoming 101 Rejections in Computer Networking Technology Classes," *IPWatchdog*, September 3, 2021, https://

ipwatchdog.com/2021/09/03/its-all-in-the-hardware-overcoming-101-rejections-in-computer-networking-technology-classes/id=137300/.

21. Kenneth Silverman, *Lightning Man: The Accursed Life of Samuel F. B. Morse* (Da Capo, 2004), 289.

22. Gerardo Con Diaz, "The Long History of Software Patenting in the United States," in Stephen H. Haber and Naomi R. Lamoreaux, eds., *The Battle over Patents: History and Politics of Innovation* (Oxford Univ., 2021), 283–03.

23. Yonover and Crowe, 94–96. The discussion regarding novelty, utility, and obviousness draws on Yonover and Crowe, chap. 10, pp. 92–102, as well as perusal of various articles linked at the end of Quinn, "How to Write a Patent Application," and unpublished notes of the author's.

24. IPO, "Comparison of Selected Sections of Pre-AIA and AIA U.S. Patent Law."

25. Gene Quinn, "Tricks & Tips to Describe an Invention in a Patent Application," *IPWatchdog*, December 26, 2015, https://ipwatchdog.com/2015/12/26/tricks-tips-for-describe-an-invention-in-a-patent-application-2/id=64133/.

26. Randall Rader, "Rader's Ruminations–Patent Eligibility II: How the Supreme Court Ignored Statute and Revived Its Innovation-Killing Two-Step," *IPWatchdog*, March 25, 2024, https://ipwatchdog.com/2024/03/24/raders-ruminations-patent-eligibility-ii-supreme-court-ignored-statute/id=174539/.

27. See "Pre-AIA §102" in Gene Quinn, "Patentability: The Novelty Requirement of 35 U.S.C. 102," *IPWatchdog*, June 10, 2017, https://ipwatchdog.com/2017/06/10/patentability-novelty-requirement-102/id=84321/.

28. See "The Utility Requirement–35 U.S.C. § 101" in Gene Quinn, "Patentability Overview: When Can an Invention Be Patented?" *IPWatchdog*, June 3, 2017, https://ipwatchdog.com/2017/06/03/patentability-invention-patented/id=84071/.

29. See Gene Quinn, "Patentability: The Nonobviousness Requirement of 35 U.S.C. 103," *IPWatchdog*, June 17, 2017, https://ipwatchdog.com/2017/06/17/patentability-nonobviousness-35-usc-103/id=84716/.

30. Khan, "History Matters," 348–49.

31. *Evans v. Jordan & Morehead*, 13 U.S. (9 Cranch) 199 (1815). For context, see this case cited in Adam Mossoff, "The Constitutional Protection of Intellectual Property," Heritage Foundation Report, March 8, 2021, https://www.heritage.org/economic-and-property-rights/report/the-constitutional-protection-intellectual-property.

32. Choate, *Hot Property*, 243.

33. Henry Campbell Black, *Black's Law Dictionary*, 6th ed. (West Publishing, 1990), "Willful," 1599.

34. B. Zorina Khan, "Trolls and Other Patent Inventions: Economic History and the Patent Controversy in the Twenty-First Century," National Bureau of Economic Research draft paper, September 2013, https://sls.gmu.edu/cpip/wp-content/uploads/sites/31/2013/09/Khan-Zorina-Patent-Controversy-in-the-21st-Century.pdf?_gl=1*3k6ilx*_ga*MTc2NDA4ODgyMC4xNzExMTQ0MTU0*_ga_N7C0HXWJJQ*MTcxMTE0NDE1NC4xLjAuMTcxMTE0NDE1NC42MC4wLjA (accessed March 22, 2024), 12.

35. Silverman, *Lightning Man*, 262.

36. Adam Mossoff, "Patent Licensing and Secondary Markets in the Nineteenth Century," *George Mason Law Review* 22 (2015): 960–61, https://papers.ssrn.com/sol3/papers.cfm?abstract_id=2602902.

37. Khan and Sokoloff ("Lives of Invention," 191) note how patents as intangible assets benefitted nineteenth-century inventors: "Inventors who chose to realize the fruits of their technological creativity through direct exploitation (a business enterprise focusing on production) might not seem to have been so affected by the patent system, but in fact even this group benefited. They were obviously helped by holding a monopoly on the use of the respective technology, but many of them were also aided in mobilizing capital for their firms by being able to report patents (or contracts committing patents granted in the future) as assets. Patent portfolios were especially useful as a signal for those who wished to attract venture capital for exceptionally innovative projects that might otherwise have seemed overly risky."

38. Khan and Sokoloff, "Lives of invention," 189.

39. Jill Jonnes, *Empires of Light: Edison, Tesla, Westinghouse, and the Race to Electrify the World* (Random House, 2004), 337.

40. Beauchamp, in "Dousing the Fires of Patent Litigation," says, "Most patents are never litigated; most patent owners never sue" (p. 137). Nevertheless, he characterizes the 1840s to 1880s American patent system as "extremely litigious" (p. 136).

41. Adam Mossoff, "The Rise and Fall of the First American Patent Thicket: The Sewing Machine War of the 1850s," *Arizona Law Review*, vol. 53 (2011): 165–211, https://papers.ssrn.com/sol3/papers.cfm?abstract_id=1354849. For litigation rate statistics, see Adam Mossoff, "The 'Patent Litigation Explosion' Canard," Truth on the Market, October 18, 2012, https://truthonthemarket.com/2012/10/18/the-patent-litigation-explosion-canard/.

42. Khan and Sokoloff, "Lives of Invention," 192–93.

43. Mossoff, "The Rise and Fall of the First American Patent Thicket."

44. Beauchamp, "Dousing the Fires of Patent Litigation," 149–53.

45. *Black's Law Dictionary* defines "Injunction" as "a court order prohibiting someone from doing some specified act or commanding someone to undo some wrong or injury. . . . Generally, it is a preventive and protective remedy, aimed at future acts, and is not intended to redress past wrongs" (p. 784).

46. Silverman, *Lightning Man*, 323. Also, see Adam Mossoff, "*O'Reilly v. Morse* and Claiming a 'Principle' in Antebellum Era Patent Law," *Case Western Law Review* 71 (2020): 735–76, https://scholarlycommons.law.case.edu/caselrev/vol71/iss2/12/.

47. Ailerons are movable parts on the edges of fixed airplane wings, used for controlling an airplane's vertical direction. Rudders are similar movable parts on the upright tail of an airplane, used to steer horizontally.

48. Howard, *Wilbur and Orville*, 393. See also Walter J. Boyne, *The Smithsonian Book of Flight* (Smithsonian, 1987), 62.

49. Howard, *Wilbur and Orville*, 383–85. The final resolution of this bitter patent litigation came not in court, but in settlement due to urgent circumstances. By 1916, more than one hundred aeronautical U.S. patents had been granted. As the United States was about to enter World War I, the Army and Navy needed airplanes, but the prospect of infringement litigation and royalty agreements slowed American plane manufacturing. The military brokered an agreement among the aeronautical

patent owners to pool their patents and through the patent pool, the Manufacturers Aircraft Association, to cross-license their patents and receive payment based on the relative value of their patents. Howard, 405–6.

An important contribution to the matter of the government's intervention that effectively forced eminent domain in the form of airplane patent pooling comes from Ron D. Katznelson and John Howells. Katznelson and Howells find no historical evidence to substantiate the government's assertion of patent hold-up by aircraft patent owners. Rather, they say, the compulsory licensing move was designed to favor the government, handing it monopsony power (i.e., the sole buyer), which resulted in low royalty rates for patent owners forced to take what the government decided it would pay. See Katznelson and Howells, "The Myth of the Early Aviation Patent Hold-Up – How a U.S. Government Monopsony Commandeered Pioneer Airplane Patents," *Industrial and Corporate Change* 24, no.1 (September 14, 2013): 1–64, (2015), doi .org/10.1093/icc/dtu003, available at SSRN: https://ssrn.com/abstract=2355673.

50. Jonathan M. Barnett, "The Great Patent Grab," in Stephen H. Haber and Naomi R. Lamoreaux, eds., *The Battle over Patents: History and Politics of Innovation* (Oxford Univ., 2021), 209.

51. James Edwards, *Property Rights: The Key to National Wealth and National Security; Restoring "Morning in America" to Regain Industrial Competitiveness,* Conservatives for Property Rights, February 2018, https://www.property-rts.org/_files/ugd/651e0c_b3442c0f8fc04c36953625f78d5e440e.pdf.

52. Other positive steps to strengthen or protect the integrity of patents and IP as property rights occurred in the 1980s, for which space doesn't allow discussion. For example, the Drug Price Competition and Patent Term Restoration Act of 1984, also known as the Hatch-Waxman Amendments, simplified the approval process to speed safe and effective generic medicines to market once the brand drug's patents expired. This was coupled with up to five more years of brand drug patent term to make up for term lost to drug development and Food and Drug Administration (FDA) approval as well as potentially three to five years of regulatory exclusivity for certain new drugs (see Frederick R. Ball and Carolyn A. Alenci, "Generic Drugs: ANDAs, Section 505(b)(2) Application, Patents, and Exclusivities," in David G. Adams, Richard M. Cooper, Martin J. Hahn, and Jonathan S. Kahan, eds., *Food and Drug Law and Regulation*, 3rd ed. (Food and Drug Law Institute, 2015), 375–91.

Also, President Reagan's Commission on Industrial Competitiveness, launched in 1983, was tasked, among other things, with focusing on intellectual property's role in spurring and sustaining U.S. industrial competitiveness. "This nation's greatest competitive advantage in the past [was] ideas that helped America grow. We need to put the power of ideas to use again, for the good of our future. America needs her best minds to create technologies that will enhance America's economic leadership in the 1980's. To sustain high rates of real economic growth, we must continue to create new 'miracles' of high technology—miracles both for innovation and for modernization of the major areas of our economy in manufacturing, agriculture, and services" (Ronald Reagan, Statement on Establishment of the President's Commission on Industrial Competitiveness Online, Gerhard Peters and John T. Woolley, *The American Presidency Project*, August 4, 1983, https://www.presidency.ucsb.edu/node/262990). The

Young Commission's Committee on Research, Development, and Manufacturing laid important groundwork for U.S. IP advances, including some discussed in this chapter.

53. See Edwards, *Property Rights*, 9.

54. Jonathan M. Barnett, "The Great Patent Grab," in Stephen H. Haber and Naomi R. Lamoreaux, eds., *The Battle over Patents: History and Politics of Innovation* (Oxford Univ., 2021), 223–24.

55. For more information about the Bayh-Dole Act, see the Bayh-Dole Coalition's "Digital Library," https://bayhdolecoalition.org/digital-library/.

56. President Ronald Reagan, Executive Order 12591, "Facilitating Access to Science and Technology," April 10, 1987 (52 *Federal Register* 13414), https://bayhdolecoalition.org/wp-content/uploads/2023/05/Reagan-Executive-Order-on-Bayh-Dole-and-Federal-Tech-Transfer-Act.pdf.

57. "Driving the Innovation Economy," AUTM, https://autm.net/AUTM/media/Surveys-Tools/Documents/AUTM-Infographic-22-for-uploading.pdf (accessed March 29, 2024).

58. "About the Court," U.S. Court Appeals for the Federal Circuit, https://cafc.uscourts.gov/home/the-court/about-the-court/ (accessed March 30, 2024).

59. "Court Jurisdiction," U.S. Court of Appeals for the Federal Circuit, https://cafc.uscourts.gov/home/the-court/about-the-court/court-jurisdiction/ (accessed March 30, 2024).

60. Haber and Lamoreaux, "Introduction," 15.

61. George C. Beighley Jr., "The Court of Appeals for the Federal Circuit: Has It Fulfilled Congressional Expectations?," *Fordham Intellectual Property Media & Entertainment Law Journal* 21, no. 3 (2011): 17n, 677–78, https://ir.lawnet.fordham.edu/iplj/vol21/iss3/4.

62. Beighley, "The Court of Appeals for the Federal Circuit," 17n, 674.

63. Beighley, "The Court of Appeals for the Federal Circuit," 706–7. *En banc* means all of a court's judges participate in a case.

64. Beighley, "The Court of Appeals for the Federal Circuit," 708.

65. 52 F.3d 967 (Fed. Cir. 1995).

66. Beighley, "The Court of Appeals for the Federal Circuit," 719–20.

67. 49 F.3d 1368 (Fed. Cir. 1998).

68. Beighley, "The Court of Appeals for the Federal Circuit," 720–22.

69. See Alexander Galetovic, "Patents in the History of the Semiconductor Industry," in Haber and Lamoreaux, eds., *The Battle over Patents*, 35–38.

70. Galetovic, "Patents in the History of the Semiconductor Industry," 37–38.

71. The World Intellectual Property Organization (WIPO) and the United Nations Conference on Trade and Development (UNCTAD).

72. Committee on Research, Development, and Manufacturing, "Preserving America's International Competitiveness: A Special Report on the Protection of Intellectual Property Rights," October 1984, app. D of the President's Commission on Industrial Competitiveness, *Global Competition: The New Reality* 2 (January 1985): 339, https://babel.hathitrust.org/cgi/pt?id=pur1.32754078799537&view=1up&seq=71.

73. See Edwards, *Property Rights*, 19–20.

74. Young Commission Committee on Research, Development, and Manufacturing, "Preserving America's International Competitiveness," 347.

75. Choate, *Hot Property*, 232–33. For more in-depth discussion of the TRIPS Agreement, see Choate, *Hot Property*, 215–37, and Jayashree Watal and Antony Taubman, eds., *The Making of the TRIPS Agreement: Personal Insights from the Uruguay Round Negotiations* (World Trade Organization, 2015), https://www.wto-ilibrary.org/content/books/9789287042330/read (accessed April 1, 2024).

76. Choate, *Hot Property*, 215, 243.

77. Susan Armstrong, personal interview.

78. Beauchamp, "Dousing the Fires of Patent Litigation," 166.

Chapter 16: A Divide Between One's Creations and Secure Ownership

1. George Santyana, *The Life of Reason* (Charles Scribner's Sons, 1906), 284, https://www.google.com/books/edition/The_Life_of_Reason_Or_The_Phases_of_Huma/sCYOer5Ttn8C?hl=en&gbpv=1&dq=Those+who+cannot+remember+the+past+are+condemned+to+repeat+it.&pg=PA284&printsec=frontcover (accessed April 20, 2024).

2. "The Trouble with Patent-Troll-Hunting," *The Economist*, December 14, 2019, https://www.economist.com/business/2019/12/14/the-trouble-with-patent-troll-hunting.

3. Pat Choate, *Hot Property: The Stealing of Ideas in an Age of Globalization* (Alfred A. Knopf, 2005), 37; Alan and Ann Rothschild, *Make: Inventing a Better Mousetrap* (Maker Media, 2016), 6.

4. The Protestant Reformation of the 1500s and 1600s pushed back against corruption in the Roman Catholic Church. The Reformation drove a return to Scripture itself and the priesthood of the believer, rather than taking the pronouncements of the Pope (a fallible human being) as God's word.

5. Francis A. Schaeffer, "Escape from Reason," in *The Complete Works of Francis A. Schaeffer: A Christian Worldview*, vol. 1 (Crossway, 1982), 227–29.

6. Francis A. Schaeffer, "The God Who Is There," in *The Complete Works of Francis A. Schaeffer: A Christian Worldview*, vol. 1 (Crossway, 1982), 43.

7. Schaeffer, "Escape From Reason," 232–34.

8. For more extensive discussion of humanity's slouch toward fully embracing this line of philosophy, see Francis Schaeffer's *The God Who Is There* and *Escape from Reason*.

9. Schaeffer, *The God Who Is There*, 35.

10. "Debussy, Claude Achille," Rupert Hughes, comp., Deems Taylor and Russell Kerr, eds., Biographical Dictionary of Musicians, *Music Lovers' Encyclopedia* (Garden City, 1947), 119–20.

11. "Musical Periods: The History of Classical Music," *Musicnotes* (blog), December 4, 2018, https://www.musicnotes.com/blog/musical-periods-the-history-of-classical-music/.

12. Nancy Pearcey, *Saving Leonardo: A Call to Resist the Secular Assault on Mind, Morals, & Meaning* (B&H Publishing, 2010), 75.

13. "About Alma," Alma Deutscher, excerpt from *New York Times*, June 2019, https://www.almadeutscher.com/about (accessed August 9, 2024).

14. Schaeffer, *Escape from Reason*, 229.

15. See Charles R. Morris, *The Tycoons: How Andrew Carnegie, John D. Rockefeller, Jay Gould, and J. P. Morgan Invented the American Supereconomy* (Henry Holt, 2005) for an insightful account of the industrial and financial giants of the Gilded Age and the turn of the century, and how they built complex corporate entities that expanded the U.S. economy into a powerful industrial and innovation engine.

16. Trusts combine enterprises to gain market power, pursuing vertical and horizontal integration strategies through mergers and acquisitions (M&A). Vertical integration achieves control of the stages of an industry from product development to manufacturing to distribution and sales. Vertical mergers increase efficiency and lower costs. Horizontal integration entails combining competitors in the same line of business. Both reduce competition to some degree within a given market. Both strategies may also benefit consumers more than enough to offset the diminished competition. See Robert H. Bork, *The Antitrust Paradox: A Policy at War with Itself* (Bork Publishing, 2021), chaps. 10 and 11, for more on vertical and horizontal mergers. See Morris, *The Tycoons*, 193–94, for discussion of the Standard Oil Trust.

17. William J. Bennett, *America: The Last Best Hope*, vol. 1; *From the Age of Discovery to a World at War, 1492–1914* (Thomas Nelson, 2006), 444–45.

18. See Morris, *The Tycoons*, 216 ff. on turn-of-the-twentieth-century efforts to break up trusts.

19. Bork, *The Antitrust Paradox,* 16, 43–44.

20. Morris, *The Tycoons*, 296–99.

21. Morris, *The Tycoons*, 297.

22. See Clarence B. Carson, *A Basic History of the United States*, vol. 5: *The Welfare State, 1929–1985* (American Textbook Committee, 1990), 10–14; Olivia B. Waxman, "What Caused the Stock Market Crash of 1929—And What We Still Get Wrong About It," *Time*, October 24, 2019, https://time.com/5707876/1929-wall-street-crash/.

23. Carson, *The Welfare State, 1929–1985*, 32.

24. Carson, *The Welfare State, 1929–1985*, 35.

25. Carson, *The Welfare State, 1929–1985*, 34–38.

26. Collectivism comes in various forms, such as communism and socialism.

27. Bork, *The Antitrust Paradox*, 309.

28. For a more nuanced view of IP exclusivity and antitrust, see Thomas M. Jorde and David J. Teece, eds., *Antitrust, Innovation, and Competitiveness* (Oxford Univ., 1992), 5: "As Schumpeter (1942) suggested . . . the kind of competition embedded in standard microeconomic analysis may not be the kind of competition that really matters if enhancing economic welfare is the goal of antitrust. Rather, it is dynamic competition propelled by the introduction of new products and new processes that really counts."

29. Bork, *The Antitrust Paradox*, 436.

30. Jonathan M. Barnett, "The Great Patent Grab," in Stephen H. Haber and Naomi R. Lamoreaux, eds., *The Battle over Patents: History and Politics of Innovation* (Oxford Univ., 2021), 208.

31. Barnett, "The Great Patent Grab," 211.

32. In antitrust, one of two rules apply when assessing anticompetitive conduct: *per se* illegality or the *rule of reason*. *Per se* is Latin for "by itself." Conduct that is per

se illegal need only be shown to have taken place, such as naked restraint of trade, e.g., price fixing, competitors agreeing to charge the same price for a good or service. The rule of reason requires weighing all the factors of restrained competition, proving actual harm to competition or unreasonable anticompetitiveness in a set of circumstances. See *Black's Law Dictionary*, 1142, 1332; and *The Antitrust Paradox*, 14–15.

33. Barnett, "The Great Patent Grab," 211.
34. Barnett, "The Great Patent Grab," 211–14.
35. Barnett, "The Great Patent Grab," 210.
36. Barnett, "The Great Patent Grab," 214.
37. Barnett, "The Great Patent Grab," 217, 218.
38. See Barnett, "The Great Patent Grab," 219–28, for more detailed discussion.
39. See Barnett, "The Great Patent Grab," 228–44, for more detailed discussion.
40. Judge Bork's *The Antitrust Paradox*, first published in 1978, provides the most in-depth, articulate, influential treatment of how courts went wrong in antitrust jurisprudence and the solution known as the Consumer Welfare Standard. This standard is based on economic evidence rather than subjective factors. Consumer welfare as the guiding light is "a neutral principle for administering the antitrust laws which evaluates competitive effects and focuses on consumer welfare and whether there is harm to competition" (Bork, *The Antitrust Paradox*, 2021 ed., xix).
41. "Morning in America" was the slogan of the 1984 Reagan-Bush re-election campaign, in which the GOP ticket won in a landslide.
42. Choate, *Hot Property*, 234.
43. Choate, *Hot Property*, 245.
44. Choate, *Hot Property*, 242–45. Choate names President Lyndon B. Johnson's Presidential Commission on the Patent System and President George H.W. Bush's Advisory Committee on Patent Law Reform (p. 245). Similar proposals appear in an undated U.S. Business and Industry Council (USBIC) briefing book, "Patents, Innovation and Jobs: A Congressional Briefing: 'Harmonizing' the U.S. Patent System with Europe and Japan; The Effect on the Small Inventor," n.d.
45. Choate, *Hot Property*, 242; USBIC, "Patents, Innovation and Jobs," tab 1, "The Harmonization Agenda."
46. Martin, ed., *Phyllis Schlafly Speaks*, 152–53.
47. See James Edwards, "The Steady Separation of Patents and Private Property Rights," *Insight IEEE-USA*, April 17, 2020, https://insight.ieeeusa.org/articles/the-steady-separation-of-patents-and-private-property-rights/.
48. Ron Katznelson, "Pre-Grant Publication—The Perilous Deviation from the Patent Bargain that Causes Long Patent Application Pendencies," *IPWatchdog*, October 25, 2016, https://ipwatchdog.com/2016/10/25/pre-grant-publication-perilous-deviation-patent-bargain-causes-long-patent-application-pendencies/id=74173/.
49. See Choate, *Hot Property*, 254–58.
50. Public Law 112–29. For an in-depth look at patent legislation from the America Invents Act in 2011 and 2017, see Andrew S. Baluch, *Patent Reform: A Comprehensive Guide to Patent Reform Developments in Congress, the Executive Branch, the Courts, and the States*, 2017 ed. (Thomson Reuters, 2017).

51. See James Edwards, "The Covered Business Methods Program Must Finally Be Laid to Rest," *IPWatchdog*, August 10, 2020, https://ipwatchdog.com/2020/08/10/covered-business-method-program-must-finally-laid-rest/id=123980/.

52. Jan Wolfe, "U.S. Patent Review Board Becomes Conservative Target," *Reuters*, November 20, 2017, https://www.reuters.com/article/technology/us-patent-review-board-becomes-conservative-target-idUSL1N1NN1SI/.

53. For a more detailed discussion of PTAB and its stark differences from judicial branch proceedings, see sections II and III of Alden Abbott et al., "Crippling the Innovation Economy: Regulatory Overreach at the Patent Office," Regulatory Transparency Project, Federalist Society, August 14, 2017, https://regproject.org/wp-content/uploads/RTP-Intellectual-Property-Working-Group-Paper.pdf.

54. USIJ's white paper documents how PTAB operates at variance from the statutory language of the AIA and its legislative history. See Alliance of U.S. Startups and Inventors for Jobs (USIJ), "How 'One Bite at the Apple' Became Serial Attacks on High Quality Patents at the PTAB," October 17, 2018, https://static1.squarespace.com/static/5746149f86db43995675b6bb/t/5bd3757af9619a5ed812cb69/1540584826664/FINAL+USIJ+Serial+IPR+White+Paper+--+Oct+17+20181.pdf.

55. Gene Quinn and Steve Brachmann, "Patent Killing Fields of the PTAB: Erasing Federal District Court Verdicts on Patent Validity," *IPWatchdog*, January 14, 2018, https://ipwatchdog.com/2018/01/14/patent-killing-fields-ptab-erasing-federal-district-court-verdicts-patent-validity/id=92375/.

56. James Edwards, "The Bottom Line on Trump's PTO: Michelle Lee Must Go," *IP Watchdog*, January 24, 2017, https://ipwatchdog.com/2017/01/24/trumps-pto-michelle-lee-must-go/id=77539/.

57. Edwards, "The Steady Separation of Patents and Private Property Rights."

58. Gene Quinn, "What Is the Point of Examination If Patents Are Not Presumed Valid?" *IPWatchdog*, March 29, 2016, https://ipwatchdog.com/2016/03/29/examination-patents-not-presumed-valid/id=67556/.

59. Peter Harter and Gene Quinn, "How IPR Gang Tackling Distorts PTAB Statistics," *IPWatchdog*, April 5, 2017, https://ipwatchdog.com/2017/04/05/ipr-gang-tackling-distorts-ptab-statistics/id=81816/; Steve Brachmann and Gene Quinn, "Are More Than 90 Percent of Patents Challenged at the PTAB Defective?" *IPWatchdog*, June 14, 2017, https://ipwatchdog.com/2017/06/14/90-percent-patents-challenged-ptab-defective/id=84343/.

60. Josh Malone, "PTAB Institution Data Analysis Proves That Reforms Have Failed," *IPWatchdog*, May 21, 2020, https://ipwatchdog.com/2020/05/21/ptab-institution-data-analysis-proves-reforms-failed/id=121440/.

61. Christopher T. Zirpoli and Kevin J. Hickey, "The Patent Trial and Appeal Board and Inter Partes Review," CRS Report, no. R48016, April 8, 2024, https://sgp.fas.org/crs/misc/R48016.pdf, 18.

62. John R. Allison, Mark A. Lemley, and David L. Schwartz, "Our Divided Patent System," *Univ. of Chicago Law Review* 82, no. 3 (Summer 2015): 1100. Also, "Stanford Public Law Working Paper," no. 2510004, "Chicago-Kent College of Law Research Paper," no. 2014–28, https://ssrn.com/abstract=2510004 (accessed April 26, 2024).

63. Abbott et al., "Crippling the Innovation Economy," 18.

64. Steve Brachmann and Gene Quinn, "58 Patents Upheld in District Court Invalidated by PTAB on Same Grounds," *IPWatchdog*, January 8, 2018, https://ipwatchdog.com/2018/01/08/58-patents-upheld-district-court-invalidated-ptab/id=91902/.

65. James Edwards, "No Quiet Title for Innovators," *Economic Standard*, May 24, 2023, https://theeconomicstandard.com/no-quiet-title-for-innovators/.

66. APJs' freedom from meaningful ethical standards is widely viewed as unfair and unacceptable because of their role in allowing patent infringers and their allies to game PTAB. The cases of Centripetal Networks and VLSI, for example, illustrate the conflicts of interest APJs skirt without consequences. See Gene Quinn, "Vidal's Solution to OpenSky Abuse Encourages PTAB Extortion," *IPWatchdog*, October 5, 2022, https://ipwatchdog.com/2022/10/05/vidals-solution-opensky-abuse-encourages-ptab-extortion/id=151882/; Eileen McDermott, "PTAB Judge Who Owns Cisco Stock Withdraws from IPR Following Centripetal Claims of Bias," *IPWatchdog*, January 6, 2023, https://ipwatchdog.com/2023/01/06/ptab-judge-owns-cisco-stock-withdraws-ipr-following-centripetal-motion-claims-bias/id=154947/; and Gene Quinn, "More PTAB Conflicts: APJ Margolies Once Again Assigned to Apple Petitions," *IPWatchdog*, August 18, 2019, https://ipwatchdog.com/2019/08/18/ptab-conflicts-apj-margolies-assigned-apple-petitions/id=112275/.

67. Illustrating the importance of Mr. Iancu's administrative reforms at PTO, particularly to PTAB, a letter from approximately 500 inventors, pro-patent organizations, universities, associations, and others to U.S. Senate and House Judiciary Committee leaders expresses several reasons https://www.property-rts.org/_files/ugd/651e0c_0d3ca3605afd48e2939f347e2ce490ce.pdf (accessed April 27, 2024).

68. See Steve Brachmann, "Jim Jordan Letter to Vidal on *West Virginia v. EPA* Could Implicate USPTO's Section 101 Subject Matter Eligibility Guidelines," *IPWatchdog*, November 3, 2022, https://ipwatchdog.com/2022/11/03/jim-jordan-letter-vidal-west-virginia-v-epa-implicate-usptos-section-101-subject-matter-eligibility-guidelines/id=152538/; Eileen McDermott, "Vidal Tells Senate IP Subcommittee There Will Be Movement on ANPRM Proposals Soon," *IPWatchdog*, July 26, 2023, https://ipwatchdog.com/2023/07/26/vidal-tells-senate-ip-subcommittee-will-movement-anprm-proposals-soon/id=164257/.

69. Choate, *Hot Property*, 243–44.

70. 149 F.3d 1368, cert. denied, 525 U.S. 1093 (1999).

71. See Edwards, "The Steady Separation of Patents and Private Property Rights," and Choate, *Hot Property*, 244.

72. Nicholas R. Mattingly, "Prior User Rights and the Incentive to Keep Innovations Secret," *IPWatchdog*, March 1, 2012, https://ipwatchdog.com/2012/03/01/prior-user-rights-and-the-incentive-to-keep-innovations-secret/id=22490/.

73. Nicholas R. Mattingly, "Prior User Rights: Rewarding Those Who Don't Contribute," *IPWatchdog*, March 11, 2012, https://ipwatchdog.com/2012/03/11/prior-user-rights-rewarding-those-who-dont-contribute/id=22742/.

74. Other questions arise in connection with prior user rights and novelty. For instance, Ron Katznelson asks, "Does the U.S. Constitution empower Congress to grant patents to inventors on their inventions after they have had an unlimited period of exclusive commercial use of the invention?" See Katznelson, "The America Invents

Act May Be Constitutionally Infirm If It Repeals the Bar Against Patenting After Secret Commercial Use," *Federalist Society Review* 13, issue 3 (February 1, 2013), https://fedsoc.org/fedsoc-review/the-america-invents-act-may-be-constitutionally-infirm-if-it-repeals-the-bar-against-patenting-after-secret-commercial-use.

75. James Edwards, "The FTC Should Give Up Its Doomed Fight with Qualcomm and Adopt Delrahim's New Madison Approach," *IPWatchdog*, September 29, 2020, https://ipwatchdog.com/2020/09/29/ftc-give-doomed-fight-qualcomm-adopt-delrahims-new-madison-approach/id=125713/.

76. See U.S. Department of Justice Antitrust Division, "United States' Statement of Interest Concerning Qualcomm's Motion for Partial Stay of Injunction Pending Appeal," U.S. Court of Appeals for the Ninth District in re: *Federal Trade Commission v. Qualcomm*, July 16, 2019, https://www.justice.gov/d9/case-documents/attachments/2019/07/16/369345.pdf.

77. James Edwards, "Order of the New Day: IP Rights in Dynamic Competition," *IPWatchdog*, June 10, 2018, https://ipwatchdog.com/2018/06/10/order-new-day-ip-rights-dynamic-competition/id=98212/.

78. No. 19-16122 (9th Cir. 2020).

79. James Edwards, "Box Score on the New Madison Approach to Antitrust and Patents," *IPWatchdog*, February 9, 2021, https://ipwatchdog.com/2021/02/09/box-score-new-madison-approach-antitrust-patents/id=129824/.

80. See Adam J. White, "The Power Broke Her," *Commentary*, March 2024, https://www.commentary.org/articles/adam-white/ftc-lina-khan-regulatory-uncertainty-power/.

81. Molly Ball and Brody Mullins, "Biden's Trustbuster Draws Unlikely Fans: 'Khanservative' Republicans," *Wall Street Journal*, March 25, 2024, https://www.wsj.com/politics/policy/lina-khan-ftc-antitrust-khanservatives-a6852a8f?st=66n3ng6peuegn3n&reflink=desktopwebshare_permalink.

82. See President Joseph R. Biden Jr., "Executive Order on Promoting Competition in the American Economy," July 9, 2021, https://www.law.berkeley.edu/wp-content/uploads/2022/04/Executive-Order-on-Promoting-Competition-in-the-American-Economy-_-The-White-House.pdf .

83. Rachel Chiu, "Big Tech's Critics Forget Government's Role in the Gilded Age," *National Review*, November 23, 2021, https://www.nationalreview.com/2021/11/big-techs-critics-forget-governments-role-in-the-gilded-age/.

84. A notable exception occurred after the Biden administration withdrew the 2019 joint policy statement on antitrust remedies available to standard-essential patents (SEPs). It floated a draft replacement statement of a one-sided, heavily antitrust, weak-IP nature, then after significant public opposition, it withdrew the draft, leaving SEPs in a case-by-case position. See Conservatives for Property Rights' comments (February 1, 2022, https://www.property-rts.org/_files/ugd/651e0c_020753474d9c4d7a9664f629f6b9fe8b.pdf); and James Edwards, "DOJ Withdraws Proposed Threat to National Security," *RealClear Policy*, August 4, 2022, https://www.realclearpolicy.com/articles/2022/08/04/doj_withdraws_proposed_threat_to_national_security_846083.html.

85. James Edwards, "Biden's Assault on Property Rights Is an Odd Way to Boost Competition," *RealClear Markets*, July 20, 2021, https://www.realclearmarkets.com/

articles/2021/07/20/bidens_assault_on_property_rights_is_an_odd_way_to_boost_competition_786158.html.

86. 447 U.S. 303 (1980).

87. See Paul R. Michel and Matthew J. Dowd, "From a Strong Property Right to a Fickle Government Franchise: The Transformation of the U.S. Patent System in 15 Years," *Drake Law Review* 69, no. 1 (1st quarter 2020) repr.; and Edwards, "The Steady Separation of Patents and Private Property Rights."

88. 561 U.S. 593 (2010).

89. 566 U.S. 66 (2012).

90. 569 U.S. 12-398 (June 13, 2013).

91. 573 U.S. 208 (2014).

92. Michel and Dowd, "From a Strong Property Right to a Fickle Government Franchise," 27.

93. David J. Kappos, testimony before the U.S. Senate Subcommittee on Intellectual Property, June 4, 2019, https://www.judiciary.senate.gov/imo/media/doc/Kappos%20Testimony.pdf. GARY: This page can no longer be found. Ask author to review and provide a new link for the testimony. I couldn't find one. This is a replacement link.

94. See Kevin Madigan and Adam Mossoff, "Turning Gold to Lead: How Patent Eligibility Doctrine Is Undermining U.S. Leadership in Innovation," *George Mason Law Review* 24 (April 13, 2017): 939–60, 953–54, https://papers.ssrn.com/sol3/papers.cfm?abstract_id=2943431.

95. See Madigan and Mossoff, 955–58.

96. Adam Mossoff, "Injunctions for Patent Infringement: Historical Equity Practice Between 1790–1882," June 14, 2024, forthcoming in the *Harvard Journal of Law & Technology* (2025), http://dx.doi.org/10.2139/ssrn.4870351.

97. See James Edwards, "Restore Injunctive Relief to Keep American Innovation Alive," *RealClear Policy*, June 6, 2024, https://www.realclearpolicy.com/articles/2024/06/06/restore_injunctive_relief_to_keep_american_innovation_alive_1036349.html; James Edwards, "Injunctions Give Teeth to Property Rights," *Human Events*, July 3, 2019, https://humanevents.com/2019/07/03/injunctions-give-teeth-to-property-rights/; and Edwards, "The Steady Separation of Patents and Private Property Rights."

98. 547 U.S. 388 (2006).

99. See Michel and Dowd, "From a Strong Property Right to a Fickle Government Franchise," 18–21; Edwards, "Restore Injunctive Relief to Keep American Innovation Alive"; and Adam Mossoff, "The Injunction Function: Why It Matters to Secure Patents as Property Rights," *Capitalism Magazine*, June 3, 2021, https://www.capitalismmagazine.com/2021/06/the-injunction-function-why-it-matters-to-secure-patents-as-property-rights/.

100. Kristen J. Osenga, "The Loss of Injunctions under eBay: Evidence of the Negative Impact on the Innovation Economy," *Hudson Institute Report*, February 28, 2024, https://www.hudson.org/regulation/loss-injunctions-under-ebay-evidence-negative-impact-innovation-economy#footNote4.

101. Edwards, "Injunctions Give Teeth to Property Rights."

102. Chuck Hong, "Big Tech's Abuse of the Patent System Must End—Take It from Me, I've Fought Google over IP for Years," *Fortune*, August 9, 2024, https://fortune.com/2024/08/09/big-tech-patent-system-netlist-google-prevail-act-uspto-ptab-aia-innovation/.

103. "The Trouble with Patent-Troll-Hunting," *The Economist*.

104. See James Edwards, "Impact on Inventors of SCOTUS Cert Denial in *Centripetal Networks v. Cisco*," *FedSoc* (blog), January 17, 2023, https://fedsoc.org/commentary/fedsoc-blog/impact-on-inventors-of-scotus-cert-denial-in-centripetal-networks-v-cisco. Note: I consulted for Centripetal Networks, 2022 to 2024, in connection with its litigation, legislative, and policy strategies.

105. See James Edwards, "Alexa, What's 'Efficient Infringement'?" Conservatives for Property Rights, January 10, 2020, https://www.property-rts.org/post/alexa-what-s-efficient-infringement.

106. Aaron Tilley, "When Apple Comes Calling, 'It's the Kiss of Death'," *Wall Street Journal*, April 20, 2023, https://www.wsj.com/articles/apple-watch-patents-5b52cda0?st=euee3ilzpz8ctq2&reflink=desktopwebshare_permalink.

107. Rebecca Tapscott, "Centripetal Networks Awarded $1.9 Billion in Infringement Suit Against Cisco," *IPWatchdog*, October 9, 2020, https://ipwatchdog.com/2020/10/09/centripetal-networks-awarded-1-9-billion-infringement-suit-cisco/id=126094/. See Edwards, "Impact on Inventors of SCOTUS Cert Denial."

108. Eileen McDermott, "Federal Circuit Upholds Mixed ITC Determination Authorizing Google Redesigns," *IPWatchdog*, April 8, 2024, https://ipwatchdog.com/2024/04/08/federal-circuit-upholds-mixed-itc-determination-google-redesigns/id=175158/.

109. Diba Mohtasham, "Apple Watch Users Are Losing a Popular Health App after Court's Ruling in Patent Case," *NPR*, January 18, 2024, https://www.npr.org/2024/01/18/1225432506/apple-watch-blood-oxygen-levels-pulse-patent-masimo.

110. Wesley Hilliard, "Apple Won't License Masimo's Patents Despite Apple Watch Import Ban," *Apple Insider*, February 1, 2024, https://appleinsider.com/articles/24/02/01/apple-wont-license-masimos-patents-despite-apple-watch-import-ban.

111. See James Edwards, "A Patents as Property Rights History Lesson," *IPWatchdog*, September 7, 2017, https://ipwatchdog.com/2017/09/07/patents-property-rights-history-lesson/id=87644/.

112. 584 U.S. ___ (2018).

113. Renee C. Quinn and Steve Brachmann, "Supreme Court Issues Much Anticipated Oil States and SAS Decisions," *IPWatchdog*, April 24, 2018, https://ipwatchdog.com/2018/04/24/supreme-court-issues-much-anticipated-oil-states-sas-decisions/id=96302/.

114. Michel and Dowd, "From a Strong Property Right to a Fickle Government Franchise," 35–36.

115. President Biden, "Executive Order on Promoting Competition in the American Economy."

116. "Statement on Biden Executive Order on Market Concentration," Conservatives for Property Rights, July 9, 2021, https://www.property-rts.org/_files/ugd/651e0c_83a7dd812d294103b8e4d171a431565a.pdf.

117. Santyana, *The Life of Reason*, 284.
118. Michel and Dowd, "From a Strong Property Right to a Fickle Government Franchise," 45.

Chapter 17: Restoring the Biblically Based Creativity-Ownership Link

1. C. S. Lewis, *Mere Christianity* (Macmillan, 1952), 36.
2. Stephen Haber, "Patents and the Wealth of Nations," *George Mason Law Review* 23, no. 4 (2016): 811–35, https://papers.ssrn.com/sol3/papers.cfm?abstract_id=2776773, 815.
3. President Ronald Reagan, "America's Economic Bill of Rights," Ronald Reagan Presidential Library and Museum Archives, July 3, 1987, https://www.reaganlibrary.gov/archives/speech/americas-economic-bill-rights.
4. Discussion of the Western canon—the foundational elements that undergird American liberty and law—fall outside the scope of this book. To learn more, two excellent resources on these important subjects are Russell Kirk, *The Roots of American Order*, 3rd ed. (Regnery Gateway, 1991); and John D. Danford, *Roots of Freedom: A Primer on Modern Liberty* (ISI Books, 2004).
5. See Paul R. Michel and Matthew J. Dowd, "From a Strong Property Right to a Fickle Government Franchise: The Transformation of the U.S. Patent System in 15 Years," *Drake Law Review* 69, no. 1 (1st quarter 2020): 40.
6. Rep. Thomas Massie, at Conservatives for Property Rights briefing, "Property Rights: The Key to National Wealth and National Security," February 28, 2018, at 2325 Rayburn House Office Building, Washington, https://www.property-rts.org/_files/ugd/651e0c_bc3459d9d810420da565acd68d7ad0a3.pdf.
7. See Mark F. Schultz, "The Importance of an Effective and Reliable Patent System to Investment in Critical Technologies," Alliance of U.S. Startups & Inventors for Jobs Policy Report, July 2020, https://static1.squarespace.com/static/5746149f86db43995675b6bb/t/5f2829980ddf0c536e7132a4/1596467617939/USIJ+Full+Report_Final_2020.pdf.
8. Thomas Massie, interview by author, audio recording via telephone, May 14, 2024.
9. Michel and Dowd to some level agree with this route to renewal of patent policy. In "From a Strong Property Right to a Fickle Government Franchise," 42, they write that "the fate of improvements in the nation's patent policy is increasingly resting with Congress and the PTO."
10. Massie, personal interview.
11. Jonathan M. Barnett and David J. Kappos, "Restoring Deterrence: The Case for Enhanced Damages in a No-Injunction Patent System," 129–51, in *5G and Beyond: Intellectual Property and Competition Policy in the Internet of Things*, Jonathan M. Barnett and Sean M. O'Connor, eds. (Cambridge Univ., 2024),132, https://ssrn.com/abstract=4034791).
12. Barnett and Kappos, "Restoring Deterrence," 151.
13. Adam Mossoff, "A Brief History of Software Patents (and Why They're Valid)," *Arizona Law Review Syllabus* 56, no. 4 (August 7, 2014): 62–77, http://dx.doi.org/10.2139/ssrn.2477462.

14. Kristen Jakobsen Osenga, "Restoring Predictability to Patent Eligibility," George Mason Univ. Center for Intellectual Property x Innovation Policy Brief, April 2024; https://cip2.gmu.edu/wp-content/uploads/sites/31/2024/04/GMU-C-IP2-Osenga-PolicyBrief-PERA-FINAL-for-web.pdf?mc_cid=707122b875&mc_eid=1c1fbe002a.

15. Saurabh Vishnubhakat, "Toward the Substitutionary Promise of PTAB Review," George Mason Univ. Center for Intellectual Property x Innovation Policy Brief (April 2024; https://cip2.gmu.edu/wp-content/uploads/sites/31/2024/04/GMU-C-IP2-Vishnubhakat-PolicyBrief-PREVAIL_FINAL-WEB.pdf?mc_cid=ee9579b3f9&mc_eid=5615b36980.

16. See Vishnubhakat, "Toward the Substitutionary Promise of PTAB Review."

17. Lewis, *Mere Christianity*, 36.

18. Renee C. Quinn, "Gene Quinn Presented with Friend of American Invention Award," *IPWatchdog*, June 3, 2024, https://ipwatchdog.com/2024/06/03/gene-quinn-presented-friend-american-invention-award/id=177323/.

19. National Institute of Standards and Technology, "Draft Interagency Guidance Framework for Considering the Exercise of March-In Rights," docket no. NIST-2023-0008, December 8, 2023, https://www.federalregister.gov/documents/2023/12/08/2023-26930/request-for-information-regarding-the-draft-interagency-guidance-framework-for-considering-the.

20. David Kappos, "President Biden's March-in Framework: A Conversation with Joe Allen, Kate Hudson, and Dave Kappos," Bayh-Dole Coalition webinar, December 14, 2023, https://bayhdolecoalition.org/wp-content/uploads/2023/12/President-Bidens-March-in-Framework-A-Conversation-with-Joe-Allen-Kate-Hudson-AAU-and-Dave-Kappos-C4IP-Transcript.pdf.

21. See Richard A. Epstein and Adam Mossoff, "Hands Off Bayh-Dole: Biden Administration Should Not Kill This 'Golden Goose' of Innovation," *The Messenger*, December 27, 2023, https://www.hudson.org/hands-bayh-dole-biden-administration-should-not-kill-golden-goose-innovation-adam-mossoff.

22. For more information about Bayh-Dole's march-in provisions and their potential misuse as price controls, see Joseph Allen, "New March-In Guidelines Threaten U.S. Innovation," *IPWatchdog*, December 10, 2023, https://ipwatchdog.com/2023/12/10/new-march-guidelines-threaten-u-s-innovation/id=170491/; the Bayh-Dole Coalition's "Issue Brief: March-In Rights Under the Bayh-Dole Act," March 2023, https://bayhdolecoalition.org/wp-content/uploads/2023/02/BDC-Issue-Brief-March-in-Rights.pdf; and "Bayh Dole & Biopharmaceutical Innovation: Myths vs. Facts," https://bayhdolecoalition.org/wp-content/uploads/2021/02/Bayh-Dole-Myths-vs-Facts.pdf (accessed June 8, 2024).

23. See Alden Abbott, "U.S. Biomedical Leadership Threatened By NIH Licensing Guidelines," *Forbes*, February 7, 2025, https://www.forbes.com/sites/aldenabbott/2025/02/07/us-biomedical-leadership--threatened-by-nih-licensing-guidelines/.

24. During the debate over the America Invents Act in 2011, prolific inventor Steve Perlman told lawmakers that first-to-invent and the pre-AIA one-year grace period made an invention process possible that was "uniquely American." The Academy Award–winning inventor attributed to first-to-invent and the grace period the invention of MOVA Contour, the computer-generated technology used for facial

reverse-aging in the film *A Curious Case of Benjamin Button*. Employing these aspects of the Creativity-Ownership model, Perlman and his colleagues over five years came up with around one hundred inventions, ultimately using six of them in the final version of the computer system and patenting only those six. ". . . because the US is a 'First-to-Invent' country and so long as we carefully document each invention, we maintain priority to the date of conception." U.S. Business and Industry Council (USBIC) briefing book, "Patents, Innovation and Jobs: A Congressional Briefing: 'Harmonizing' the U.S. Patent System with Europe and Japan; The Effect on the Small Inventor," n.d., tab 5, February 28, 2011, letter from Perlman to Sen. Dianne Feinstein.

25. The PTO reports traditional patent pendency, "The average number of months from the patent application filing date to the date the application has reached final disposition," as 26.1 months in December 2024. PTO took 20.3 months on average before its first action on a new application and 29.9 months for total traditional pendency with a request for continued examination (RCE). Patent applications with at least one RCE were pending an average of 42.9 months—more than three-and-a-half years. "Patents Pendency Data December 2024," USPTO, https://www.uspto.gov/dashboard/patents/pendency.html.

26. See Bio.News Staff, "WTO Ministerial Ends, No Expansion of COVID IP Waiver," Bio.News, March 4, 2024, https://bio.news/international/wto-ministerial-covid-ip-intellectual-property-waiver-diagnostics-therapeutics/; and Eileen McDermott, "C4IP and IP Celebrities Tell Biden to Pass on Extension of TRIPS Waiver," *IP Watchdog*, December 5, 2023, https://ipwatchdog.com/2023/12/05/c4ip-ip-celebrities-tell-biden-pass-extension-trips-waiver/id=170367/.

27. President's Commission on Industrial Competitiveness Committee on Research, Development, and Manufacturing, "A Special Report on the Protection of Intellectual Property Rights," Preserving America's Industrial Competitiveness, app. D, Oct. 1984, in President's Commission on Industrial Competitiveness, *Global Competition: The New Reality*, report, vol. 1 (Washington: Government Printing Office, January 1985), 320.

28. U.S. Patent and Trademark Office, National Institute of Standards and Technology, and U.S. Department of Justice, Antitrust Division, "Policy Statement on Remedies for Standards-Essential Patents Subject to Voluntary F/RAND Commitments," December 19, 2019, https://www.justice.gov/atr/page/file/1228016/dl.

29. U.S. Department of Justice Office of Public Affairs, "Assistant Attorney General Makan Delrahim Delivers Remarks at the USC Gould School of Law's Center for Transnational Law and Business Conference," November 10, 2017, https://www.justice.gov/opa/speech/assistant-attorney-general-makan-delrahim-delivers-remarks-usc-gould-school-laws-center. Also see Kirti Gupta, "The New Era of Antitrust Law and Policy in Standards: Embracing Evidence Based Policy-making," *IPWatchdog*, November 30, 2017, https://ipwatchdog.com/2017/11/30/new-era-antitrust-law-policy-standards-embracing-evidence-based-policy-making/id=90635/.

30. President Joseph R. Biden Jr., "Executive Order on Promoting Competition in the American Economy," July 9, 2021, https://www.law.berkeley.edu/wp-content/uploads/2022/04/Executive-Order-on-Promoting-Competition-in-the-American-Economy-_-The-White-House.pdf.

31. Haber, "Patents and the Wealth of Nations," 815.

32. Lewis, *Mere Christianity*, 36.

33. James Edwards, "Weakening America's Patent System Will Allow China to Dominate Technology of the Future," *RealClear Policy*, May 4, 2021, https://www.realclearpolicy.com/articles/2021/05/04/chinas_patent_gain_is_our_loss_775512.html. See Dowd and Michel, "From a Strong Property Right to a Fickle Government Franchise," 44–49.

34. Thomas M. Jorde and David J. Teece, eds., *Antitrust, Innovation, and Competitiveness* (Oxford Univ., 1992), 5.

Chapter 18: Conclusion: Reinventing by a Biblical Return

1. Francis A. Schaeffer, "The God Who Is There," in *The Complete Works of Francis A. Schaeffer: A Christian Worldview*, vol. 1 (Crossway, 1982), 100.

2. Os Guinness, *The Call: Finding and Fulfilling the Central Purpose of Your Life* (W, 2003), 46.

3. Madeleine L'Engle, "The Expanding Universe," Newbery Medal acceptance speech, August 1963, in *A Wrinkle in Time* (Square Fish, 2007), 244–45.

4. Francis Bacon, *Novum Organum*, Joseph Devey, ed. (Collier & Son, 1902), https://archive.org/details/novumorganum0000baco/page/4/mode/2up, 290. Schaeffer also discusses Bacon, quoting this passage in Francis A. Schaeffer, "Escape from Reason," in *The Complete Works of Francis A. Schaeffer: A Christian Worldview*, vol. 1 (Crossway, 1982), 226.

5. Nancy Pearcey, *Saving Leonardo: A Call to Resist the Secular Assault on Mind, Morals, & Meaning* (B&H, 2010), 107.

6. Bacon, *Novum Organum*, Devey, ed., 105–6. See also John W. Danford, *Roots of Freedom* (ISI, 2004), 63.

7. Thomas Massie, interview by author, audio recording, May 14, 2024.

8. Massie, personal interview.

9. Paul DeRosa, "The Eclectic Brilliance of Edmund Phelps," *American Interest* 9, no. 5 (April 20, 2014).

10. DeRosa, "The Eclectic Brilliance of Edmund Phelps."

11. Phoebe Miles, "To Invent Is Divine: How Strong Patent Rights Help Us to Reflect God's Innovative Nature," with James Edwards at Friday Forum, Faith & Law, July 8, 2022, 2172 Rayburn House Office Building, Washington, DC, 08:09, https://faithandlaw.org/resources/2022/05/to-invent-is-divine/.

12. Miles, "To Invent Is Divine," 09:35.

13. McKenzie Prillaman, "Stanford Explainer: CRISPR, Gene Editing, and Beyond," *Stanford Report*, June 10, 2024, https://news.stanford.edu/stories/2024/06/stanford-explainer-crispr-gene-editing-and-beyond#questions; Dennis Normile, "Chinese Scientist Who Produced Genetically Altered Babies Sentenced to 3 Years in Jail," *Science*, December 30, 2019, https://www.science.org/content/article/chinese-scientist-who-produced-genetically-altered-babies-sentenced-3-years-jail.

14. Os Guinness, *The American Hour: A Time of Reckoning and the Once and Future Role of Faith* (Free Press, 1993), 357.

15. Guinness, *The American Hour*, 357.

16. Guinness, *The American Hour*, 358.

17. Guinness, *The American Hour*, 358, 359.
18. Massie, personal interview.
19. Schaeffer, "Escape from Reason," 220.
20. Francis A. Schaeffer, "Death in the City," in *The Complete Works of Francis A. Schaeffer: A Christian Worldview*, vol. 4 (Crossway, 1982), 298–99.

Index

3D-Predict *139*
3G *140*
4G *140*
5G *140*
1903 Wright Flyer *14, 48–51*
2019 joint SEP policy statement *249, 251*

A

abstract idea *225–26*
A Connecticut Yankee in King Arthur's Court *186*
Ada, Countess of Lovelace *42*
Adam *9, 18, 39–41, 82, 98, 110, 150, 254, 261*
Adams, John *153–54, 170, 179*
Adams, Samuel *170*
administrative patent judges (APJs) *221*
Administrative State *231, 235*
Alice Corp. v. CLS Bank *226, 241*
Alice in Wonderland *86*
Alice-Mayo Framework *226, 230, 242*
Allen, Joseph P. *204*
Alliance of U.S. Startups & Inventors for Jobs (USIJ) *237*
Allison, James P. *55, 72–73*

Amahl and the Night Visitors *212*
"Amazing Grace" *86*
Amazon *229*
America *154*
America Invents Act (AIA) *220–23, 222–23, 229, 239, 242–43, 247–48*
American Century *201*
American Inventors Protection Act *219*
American IP system *208*
American patent system *201, 208*
American Revolution *152, 169–71, 174*
American Society of Mechanical Engineers *213*
A Midsummer Night's Dream *14*
Ananias and Sapphira *116*
Android *140*
Antheil, George *7, 74*
antitrust *202, 213–18, 224, 231*
Antitrust Division *239, 249*
antitrust reform *248–50*
Apollo Block I Guidance Computer *6*
Apostles' Creed *28*
Apple *140, 228–29, 256*
applied research *51, 55–62, 236*

Aristotle 86, 149
Armstrong, Susan M. 61–62, 130–31, 208
Article I, Section 8, Clause 8 173, 175, 233–34
Articles of Confederation 152, 170, 175, 179
artificial intelligence (AI) 41–42, 262
Association for Molecular Pathology v. Myriad Genetics 226
AT&T 92
Augustine 86
AutoSyringe 47
"Ave Maria" 136
A Wrinkle in Time 86

B

Bach, Johann Sebastian 13, 86, 211
Bacon, Francis 256–59
Balancing Incentives Act 243
Barnabas 115
Barnett, Jonathan 216–17, 241
Baroque era 211–12
basic research 17, 53–55, 236
Batchelor, Charles 68, 125
Bayh, Birch 203–4, 245
Bayh-Dole Act 21, 202–4, 217, 244–46, 249, 251
Beethoven, Ludwig van 211
Belgic Confession 29
Bell, Alexander Graham 42, 185, 186, 199, 256, 262
Bell Laboratories 5–6, 92
Bell Telephone Company 92, 212
Benjamin, Miriam 185
Berkhof, Louis 38, 64

Berlin, Irving 13
Berra, Yogi 42
Bible 19, 25–30, 77, 81, 100, 105, 113, 146–49, 151, 160, 178–82, 212, 264–66
Biden administration 224, 244–46, 249, 251
Biden, Joseph 231
Big Tech 228, 248, 250
Bilski v. Kappos 225, 241
biotechnology 237, 262
Black's Law Dictionary 159, 165, 197
Blackstone, William 179
Blanchard gun-stocking lathe 53
Blanchard, Helen 185
Blanchard, Thomas 53
Bork, Robert 214–17, 231
Boston Scientific 93
Brahms, Johannes 211
British Royal Society of Arts 72
Burke, Edmund 145

C

Cade, James Robert 17, 58–59
Cade Museum for Creativity and Invention 82, 177, 259
Carson, Clarence B. 213–14
Carver, George Washington 71–72, 129, 255
Catholicism 74
Cayley, George 18
Celera 136
centralized economy 214
Centripetal Networks 222, 229
Cephus Catheter 58
Chanute, Octave 45
Chariots of Fire 142
Charles II, king 162–63

Charpentier, Emmanuelle *255*
China *142, 219, 226, 231, 246, 250, 262*
Choate, Pat *165–66, 170–71, 190, 217*
Chopin, Frederic *211*
Christendom *151*
Cisco *229*
Civil War *210*
Classical era *211*
Clayton Act *213*
Clemson University *139*
Clucks Capacitor *75*
Coca-Cola *164–65*
code division multiple access (CDMA) *7, 52*
Cold War *201–2*
Cole, Nat King *52*
collectivism *214*
Combining Faculty, the *42*
commercialization *16, 68–69, 139–40, 198, 202–03, 208, 238, 245, 248*
Commission on Industrial Competitiveness *206–7*
Common Grace *11, 19, 63–78, 83, 89, 119, 150–51, 172, 178, 182, 211, 232, 235, 238, 252, 255, 257–59, 259, 261, 264*
communism *21, 97*
Communist *262*
compulsory license *216–17*
compulsory licensing *207*
Congress *76, 190, 193, 204–5, 219–20, 225, 239–44, 248–49*
Congressional Research Service *221*
Constitutional Convention *152, 170–72*

Consumer Financial Protection Bureau *235*
Consumer Welfare Standard *217*
Continental Congress *170*
Coons, Chris *242*
copyright *21, 158–63*
Copyright Act of 1790 *174*
Copyright Act of 1976 *161*
Corelli, Arcangelo *86*
cotton gin *185*
Covered Business Methods (CBM) *220*
Creation Mandate *11, 18–19, 36–40, 66, 82–85, 110, 116, 150, 181–83, 211, 253, 256, 264*
creation of wealth *112*
creativity, human *7–11, 13–18, 81–88, 89–93*
CRISPR *262*
Crouch, Andy *19, 40–43, 80, 85, 88, 141*
crowdsourcing *54*
Crowe, Ellie *159, 192*
Curtiss, Glenn *200, 216*
cybersecurity *229*

D

Dallimore, Arnold *180*
Damadian, Raymond V. *77*
Darwinist *73*
David, king *14, 232*
da Vinci, Leonardo *14, 18, 87, 129, 255*
Davy, Humphrey *54*
Debussy, Claude *211*
Declaration of Independence *151–52, 170–72, 179*
Defend Trade Secrets Act of 2016 *166*

Delrahim, Makan 224, 249
Democracy in America 183
Department of Commerce 204
Department of Defense (DOD) 203, 224
Department of Energy 224
Department of Justice (DOJ) 222, 224, 249
Deutscher, Alma 13, 212
Dewey, John 213
Diamond v. Chakrabarty 225
Dickinson, John 20, 117
disruptive technology 210
Dole, Robert 203, 245
double-balloon catheter 58
Doudna, Jennifer 255
Douglas, Dewayne 58
Dowd, Matthew 226, 230
Dreisbach, Daniel L. 116–17, 179
dynamic competition 215, 224, 231, 248–49, 251
Dynamism 83
dynamo 55

E

Eastern Paper Bag Company 185
Eastman Kodak 212
eBay v. MercExchange 227, 240
Eckert, J. Presper 53
Economic Bill of Rights 234
economic dynamism 83–84
Economic Espionage Act of 1996 166
economic value 137–41
Edison Electric Light Company 69
Edison General Electric 69
Edison Electric 212

Edison, Thomas A. 51, 60, 68–70, 124–27, 159–60, 186, 197, 255, 262
Edwards, Jonathan 179–80
efficient infringement 227–30, 247
Eine Kleine Nachtmusik 13
electrical circuit 4
"Electrode arrangement of gas tubes" 76
electrolyte replacement 58
electromagnetism 54
Eli Lilly 93
Enlightenment 151, 153–54, 179, 210
Erler, Edward 152, 173
Europe 13, 145, 151, 154, 190, 201, 218, 231
European Patent Office 226
Eve 9, 39, 82, 98, 150, 254
executive order 12591 204
executive order 14036 249
ex nihilo 26–30, 34, 36, 41
Experimental Researches in Electricity and Magnetism 68
Extreme Spectrum 56
Ex Vivo 3D Cell Culture 139–40

F

Fairchild Semiconductor 5, 7
Fall, the 254, 256–57, 261
Faraday, Michael 54–55, 60, 68, 70
Federal Courts Improvement Act 204
Federalist Papers 153
Federalists 153
Federal Technology Transfer Act 204

Federal Trade Commission Act   *213*
Federal Trade Commission (FTC)   *224, 235, 239, 248–49*
first-to-invent   *222, 246, 251*
flat-bottomed paper sack   *185*
foliage penetration radar   *56*
Ford, Henry   *69*
Founders   *153, 171–78, 174–83, 186, 188, 191, 201, 214, 225, 230, 244, 246, 258*
Four Seasons   *13*
FRAND   *249*
Franklin, Benjamin   *170–73, 179, 186*
freedom to operate   *246–47*
free enterprise   *234–35*
Free Enterprise: A Judeo-Christian Defense   *85, 118*
"frequency-hopping"   *74*
fruits of one's labor   *11, 13–23, 39, 66, 70, 78, 80–88, 113, 119, 121, 126–28, 134–37, 150–52, 155–56, 175, 234, 249*
FTC v. Qualcomm   *224*
fuel of interest, fire of genius   *23, 123, 126, 144, 214*

G

Gable, Clark   *73*
Galaxy   *256*
Galilei, Galileo   *65*
Garden of Eden   *9, 36, 38–41, 98, 254, 261*
Gatorade   *17, 58–59, 177, 259*
GE   *69*
Genentech   *93*
General Agreement on Tariffs and Trade (GATT)   *206–7, 219*
George III, king   *117*
Germany   *201–2*
Gerry, Elbridge   *170*
Gevaert, Matthew   *139*
God's creativity   *11, 253*
God's ownership   *11, 119, 253*
God the Creator   *25–34*
God the Owner   *97–105*
golden age of American patenting   *191–201, 208–9, 211, 214–15, 215, 231, 234, 240, 250, 251*
gong and signal light device   *185*
Goodyear, Charles   *197*
Google   *140, 228–29*
Gore, Robert   *52*
Gore-Tex   *52*
Gould, Jay   *68, 212*
grace period   *247, 251*
Graham, Billy, Rev.   *77*
Great Awakening   *178–83, 210, 265*
Great Depression   *201, 213–15*
Greece   *154*
Greek   *13*
Guinness, Os   *254, 262–63*
Gutenberg, Johannes   *160*

H

Haber, Stephen   *250, 258, 260*
Hamilton, Alexander   *172, 175, 183*
Hancock, John   *170*
Handel   *13–14, 86–87, 211*
harmonization   *218–20*
Harris, Kamala   *245*
Hatch-Waxman Act   *249*
Haydn, Joseph   *211*
Hegel, Georg   *210–11*

Heidelberg Catechism 29
Henry, Matthew 26, 38, 102–5, 113
Hewlett-Packard 206
Hoerni, Jean 5
Hollywood Walk of Fame 75
Holst, Gustav 211
Holy Spirit 148, 180
Homestead Act 155
Hong, Chuck 228
Howe, Elias 197
Huckleberry Finn 86
Hughes, Howard 74
human creativity 35–45, 65–67, 76, 121–23, 212, 233, 249, 256–66
human flourishing 79–93, 112, 119, 133, 141–44, 183, 188, 201, 232–33, 251, 254, 261, 264–66
Human Genome Project 136
humanist worldview 260
human ownership 19–23, 107–19, 121–23, 134–37, 212, 233–35, 263–65
Huntington, Samuel 154
Hurricane Ian 63

I

Iancu, Andrei 222
iBot 47
IEEE-USA 77
imago Dei 8, 8–11, 26, 32, 38–40, 66, 76, 82, 89, 98–99, 110, 112, 119, 181, 182, 252–55, 255, 257, 260–61, 261, 264–66
immunotherapy 72–73
indigenous innovation 83–84
Industrial Revolution 83, 145–46

injunction 227–30, 240–41, 250
innovation ecosystem 208–9, 236–39, 251, 261
integrated circuit 5–6, 76
Intel 7
intellectual property exclusivity 215–16
intellectual property (IP) 20–23, 157–66, 173–78, 206–8, 254, 260
intellectual property rights 171, 202, 207
InterDigital 92, 140, 224
Internet of Things 90, 140
inter partes review 220–21, 230–31
invention 47–62, 262–64
invention and discovery 14–18
invention description 192
Inventivity® 82, 259
Iowa State Agricultural College 71
iPhone 140, 256
Ipilumimab 73
IP rights 230–31, 238, 251, 259
IPWatchdog.com 193, 230

L

Jacobs, Irwin 7, 47–48
Japan 201–2, 219
Jay, John 175, 179
Jefferson, Thomas 152, 170, 172–73, 195
Jesus Christ 64–66, 70, 77, 109, 113, 148, 180, 262, 264–66
Johns Hopkins University 237
Jonathan 232
Jorde, Thomas M. 251
Judeo-Christian worldview 211, 259, 263–66

Juilliard School 77
Julius Caesar 14

K

Kamen, Dean 47–48
Kant, Immanuel 210
Kappos, David 226, 241, 245
Kaptur, Marcy 243
Katznelson, Ron 76
Kendall, Amos 198
Kendall, Claire 81
Kentucky Fried Chicken 165
Khan, Zorina 190–91, 198
Kilby, Jack 3–4, 6–7, 16, 19, 255
King Lear 14
Kirk, Russell 155–56
KIYATEC 91–92, 139–40
Knight, Margaret 185
know-how 165–66
Korean War 201

L

Lafayette 111
Lamarr, Hedy 7, 73–75, 255
Landes, David S. 145–46, 149, 184
Langley, Samuel P. 45, 49–50
Lanham Act of 1946 165
law of nature 225–26
laws of nature 14, 18, 152
Lemelson Center for the Study of Invention and Innovation 15, 86
Lemelson-MIT Lifetime Achievement Award 77
L'Engle, Madeleine 86–87, 255
Lewis, C. S. 244, 250
Library of Congress 221
licensing 139
Liddell, Eric 142
Liddell, Jennie 142
light bulb 124–25
Lilienthal, Otto 18, 45, 49
Lincoln, Abraham 22, 123, 214
Lindsell, Harold 21, 85, 118, 148, 164
Little Women 86
Locke, John 135, 149–54, 172–73, 179, 255

M

MacBook Pro 161
Madigan, Kevin 226, 231
Madison, James 153, 172–75, 175, 255
Magna Carta 151
magnetic resonance imaging scanner (MRI) 77
Magnetic Telegraph Company 198
Malawer, Sid 17
march-in rights 244, 251
Markman v. Westview Instruments 205
Marshall, John 176, 196, 246
Martin guitars 13
Marxism 21, 148
Marx, Karl 92
Masimo 229
Mason, George 152, 172
Massachusetts Institute of Technology (MIT) 75, 237
Mass Flourishing 83
Massie, Thomas 59, 75–76, 237–38, 240–42, 257–59, 264
Master, Jonathan 39
Mather, Cotton 178
Mayer, Louis B. 74

Mayo Collaborative Services v. Prometheus Laboratories 225, 241
McCartney, Paul 52
McCorkle, John 56, 59, 67, 81
Medical University of South Carolina 57
Men of Progress 186–87
Menotti, Gian-Carlo 212
Messiah 13–14, 87
Metro Goldwyn Mayer (MGM) 73–74
Michelangelo 14, 86, 129
Michel, Paul 226, 230, 232
microchip 6–7, 90, 92, 130–31, 206
Miles, Phoebe Cade 59, 82, 177, 259
Modern era 211–12
Mona Lisa 14
monopolies 212
monopoly 159–61, 172
Montesquieu, Charles 179
Moore, Gordon 5
Morgan, J. P. 69
Morris, Charles R. 213
Morse, Samuel F. B. 66, 185–87, 194–95, 198–200, 255
Moses 100, 112
Mossoff, Adam 68, 173, 176, 186, 226, 231, 241
mother of invention 47, 58, 124
Mozart, Wolfgang Amadeus 13, 211–12
Musk, Elon 255
Myriad Genetics 241

N
National Academy of the Arts of Design 66
National Cancer Institute 140
National Institute of Standards and Technology (NIST) 222, 239, 244–45, 249
National Institutes of Health (NIH) 245–46
National Inventors Hall of Fame (NIHOF) 48, 72, 75, 185
Nehemiah 234
Nelson, Willie 72–73
Netlist 228
Newbery Award 86
New Deal 213–14, 217
New Madison Approach 224, 248–49, 251
Nicene Creed 28
Noah 63
Noahic covenant 63–67
Nobel Prize 5, 55, 83, 154, 258
North America 151, 154
novelty 196
Noyce, Robert N. 5–7, 19, 255
Nvidia 93

O
Obama administration 224
obviousness 196
Oil States Energy Services 230–31
Oil States Energy Services v. Greene's Energy Group 230–31, 244
one-year grace period 193, 222
open-source 175
O'Reilly, Henry 194
Original Sin 89
Osenga, Kristen 228
ownership 89–93, 97–105, 214

P

Paine, Robert Treat *179*
Palmer, Arnold *80*
Pascal, Blaise *44*
patent *21–23, 76–77, 118, 158–60, 162–64, 172, 254*
Patent Act of 1790 *174, 176–78, 178, 190*
Patent Act of 1793 *178, 190*
Patent Act of 1836 *178, 189–91*
patent assertion *199*
Patent Bargain *22, 177, 215, 223, 246*
patent claim *193–95, 200, 221–22*
patent eligibility *222, 225–27*
Patent Eligibility Restoration Act *242*
patent-eligible subject matter *194, 241–42, 250*
patent examination *194*
patent exclusivity *197*
patent infringement *197*
patent licensing *68, 197, 203–4*
patent misuse *215–16*
patent models *174, 194*
patent reform legislation *240–48*
"patent thicket" *200*
Patent Trial and Appeal Board (PTAB) *220–22, 229–31, 235, 242–44, 247–48, 250–51*
"patent trolls" *220, 227*
patent validity *200, 220–22*
Paul, the apostle *39, 64, 114–16*
Pearcey, Nancy *87, 257*
Peter, the apostle *115–16*
pharmaceuticals *237*
Phelps, Edmund *83–84, 154–55, 258–59*

Pickens, Andrew *180*
Pieta *14*
Pinckney, Charles, III *173*
Piper, John *148*
piracy *208*
planar process *5*
Plantinga, Cornelius, Jr. *43*
Plato *48, 124*
populism *213–14*
post grant review *220*
potter and the clay, the *101–2, 110–11, 114, 182*
predatory infringement *229*
presumption of validity *221, 227*
Princess and the Goblin *86*
prior art *192, 196, 219, 222, 247*
prior user rights *222, 246, 251*
Progressive Era *201*
progressivism *213–14*
Promoting and Respecting Economically Vital American Innovation Leadership (PREVAIL) Act *242*
property rights *19–23, 83–84, 89–93, 97–99, 109–12, 115–16, 117, 122, 145–56, 160, 163–66, 166, 172–78, 175–78, 182–83, 186, 189, 191, 196–99, 201, 203, 208, 212, 214–20, 222–23, 227, 230–31, 234–35, 244, 256–66, 260–61*
Protestantism *263*
Protestant Reformation *13, 151, 153–54, 160, 210–11, 265*
public domain *22, 159–61, 163–65, 177, 261*
public rights doctrine *230*
Punch magazine *188–89*

Q

Qualcomm   *7, 47–48, 52, 61–62, 93, 130–31, 140, 208, 224, 248*
quantum computing   *262*
Quinn, Gene   *193, 221, 230, 244*

R

Rader, Randall   *195*
Ravel, Maurice   *211*
RCA   *76*
Reagan, Ronald   *155, 203–4, 206, 234, 248*
Realizing Engineering, Science, and Technology Opportunities by Restoring Exclusive (RESTORE) Patent Rights Act   *240*
Reformed doctrine   *180*
Rembrandt van Rijn   *86, 129, 138*
Renaissance   *13, 151*
Republic   *124*
RescueStreamer   *60*
research and development (R&D)   *5, 17, 61, 68, 127, 176, 202, 216–17, 218, 224, 229, 245, 249, 251*
    firms   *92*
Restoring America's Leadership in Innovation Act (RALIA)   *240, 242, 244, 248*
Roaring Twenties   *201*
Rockefeller, John D.   *212*
"Rock of Ages"   *86*
Roman   *13*
Romantic era   *211*
Rome   *154*
Romeo and Juliet   *14*
Roosevelt, Franklin D.   *213–14*
Rousseau, Jean Jacques   *210*
Royal Institution   *54, 70*

Royal Society   *54*
Ruckelshaus v. Monsanto Company   *166*
Rush, Benjamin   *170, 179*

S

Salk, Jonas   *255*
Samsung   *256*
Santyana, George   *231, 246*
Saul, king   *232*
Schaeffer, Francis   *38–39, 43, 210–11, 214, 264–66*
Schlafly, Phyllis   *218*
Schubert, Franz   *136*
scientific method   *14, 17, 55*
"scope" doctrine   *215–16*
Seae Ventures   *140*
Second Treatise of Government   *149–50*
Segway   *47*
semiconductor   *3–6, 48, 130, 206, 228, 237*
sewing machine   *185*
Shakespeare, William   *14, 86*
Shark Tank (TV show)   *16*
Sherman Act   *213*
Sherman, Roger   *170, 179*
Shockley Semiconductor Laboratories   *5*
Shockley, William B.   *5–6*
Silicon Valley   *237*
Simmons, Cephus   *57, 61, 125, 141*
Simpson College   *71*
Singer, Isaac   *199*
Siskind, Lawrence J.   *159–60*
Smith, Adam   *135*
Smithsonian Institution   *45, 49, 86, 186–87*

socialism *21, 97, 148–49, 154–55*
software-implemented invention *226*
Sokoloff, Kenneth *191*
Sonos *229*
Soviet Union *262*
Spalding, Matthew *155*
standard-essential patent (SEP) *222, 224, 249*
Standard Oil *212*
Stanford University *237*
Staples, Mavis *73*
"Stardust" *52*
Starkweather, Gary *52–53*
state-owned enterprise *154–55*
State Street Bank & Trust Co. v. Signature Financial Group *205–6, 222*
Statute of Anne *161, 174*
Statute of Monopolies *174*
Steinway pianos *13*
Stevens Institute of Technology *76*
Stevenson-Wydler Technology Innovation Act *204*
Stewart, Jimmy *73*
Stokely Van Camp *59*
Stradivarius violins *13*
Stravinsky, Igor *211*
Stryker *93*

T

Tariff Act *166*
Taylor, Charlie *49*
T-cells *55, 72–73*
Tchaikovsky, Pyotr Ilyich *211*
Teece, David J. *251*
Teksler, Boris *228*
telegraph *185*

telephone *185*
Ten Commandments *89, 99–100, 108–9, 116, 146–47, 149, 182, 254*
Tennent, Gilbert *180*
Tennent, William *180*
Tesla, Nikola *128–29, 140, 210*
Texas Instruments *3–4, 7, 16*
The Last Supper *14, 87*
The Politics of Prudence *155–56*
The Secret Garden *86*
The Wealth and Poverty of Nations *145–46, 184*
Thomson, J. J. *6*
Tillis, Thom *242*
Tocqueville, Alexis de *183*
Tower of Babel *263*
Tracey, Kevin *44*
trademark *21, 158–60, 164–65*
Trade-Related Aspects of Intellectual Property Rights (TRIPS) Agreement *202, 206–8, 207, 217–19, 219, 247*
trade secret *22, 158–60, 165–66, 219*
tragedy of the commons *261*
TRIPS waiver *247*
Trump administration *224*
trusts *212*
truth, beauty, goodness *8–10, 211*
Tuskegee Institute *71, 129*
Twain, Mark *186*
Tyndall, John *60*
tyranny of numbers *3–5*

U

United States *13, 21–22, 73–74, 83, 152–53, 154, 161, 170, 174, 177, 183, 188, 190–91, 201–2,*

205–9, 217, 220, 222, 225, 231–33, 235, 237–38, 242, 247, 250, 261, 263, 265
University of California 226
University of Florida 17, 58–59
University of Pittsburgh 76
University of Texas M. D. Anderson Cancer Center 72
U.S. competitiveness 250
U.S. Constitution 14, 152, 171–78, 178–79, 183, 213
U.S. Copyright Office 158
U.S. Court of Appeals for the Federal Circuit (CAFC) 202, 204–6, 226, 229–30
useful arts 14–15, 162, 175–76, 188
U.S. International Trade Commission (ITC) 222, 229, 242, 248
U.S. mobilization 214
U.S. Patent and Trademark Office (PTO) 158, 164–65, 218–22, 239, 246, 248–49
U.S. Patent Office 186–87, 190, 196
U.S. patent system 189–208
U.S. Supreme Court 225–27, 230–31, 244
U.S. Trade Representative (USTR) 239
utility 196

V

valley of death 92, 238
Vanderbilt 69
Van Gogh, Vincent 138
Vaughn, Ellen 70
venture capital (VC) 236–38

Vermeer, Johannes 86
Vietnam War 201
vine and the fig tree, the 20, 66, 111, 116–17
Virginia Declaration of Rights 152
virtuous circle 93, 208–9, 232, 251, 260
Vivaldi, Antonio 13, 211
VLSI 222
Volvo 136

W

War Between the States 71
Washington, Booker T. 71
Washington, George 86, 111, 117, 170–72, 174
Water Music 13
Western Union 68–69
Westinghouse (company) 212
Westinghouse, George 129, 140, 186, 199, 210
Westminster Confession of Faith 26, 29
Westminster Larger Catechism 99, 108–9, 118, 146–47
West, the 83, 154, 234, 259
Whelchel, Hugh 87
Whitefield, George 178–81
Whitney, Eli 185
Wind in the Willows 86
Wired 72
wireless blood-oxygen monitor 229
wireless speakers 229
Wissolik, Erica 77
Wissolik, Raymond A. 76–77
Witherspoon, John 170, 179
work made for hire 134, 162

World Intellectual Property
    Organization (WIPO)   *52–53,
    206, 206–7, 260*
World Trade Organization (WTO)
    207, 217–19, 219, 247, 260
worldview   *262–64*
World War I   *201, 214*
World War II   *74, 201–2*
World Wide Web   *262*
Wright Brothers   *14, 18, 45, 60,
    186, 199–201, 210, 216–17,
    255, 262*
Wright v. Curtiss   *200*
Wright, Wilbur and Orville   *48–51*

<u>X</u>

Xerox   *52*
Xtandi   *245*

<u>Y</u>

"Yesterday"   *52*
Yonover, Robert   *59, 159, 192–93*
Young Commission   *206–8, 248*
Young, John A.   *206*
Yo-Yo Ma   *135–36*